Investigating Chemistry in the Laboratory

David C. Collins, Ph. D.
Colorado State University - Pueblo

W. H. Freeman and Company
New York

ISBN: 0-7167-7485-2
EAN: 97807167-7485-3

Printed in the United States of America

First printing

W. H. Freeman and Company
41 Madison Avenue
New York, NY 10010
Houndmills, Basingstoke RG21 6XS England
www.whfreeman.com

Contents

Safety in the Laboratory

Introduction

Safety in the laboratory should be first and foremost in the mind of a scientist. Accidents are common anywhere, and accidents in a chemical laboratory may have the potential of resulting in serious consequences. Common laboratory accidents include cuts, burns, contact with corrosive or toxic chemicals, and the inhalation of volatile fumes. The occurrence of each of these accidents, and others, can be minimized if simple safety precautions are followed. Although accidents can happen, you should not be afraid of the chemistry laboratory; you must experiment in order to obtain the hands-on experience necessary to understand science. If proper safety precautions are followed, the chemistry laboratory can be as safe as your own kitchen.

What follows is a brief introduction to safe laboratory practices. It is recommended that your instructor inform you of additional safety issues and the safety equipment available in your laboratory.

Eye Protection

Eye protection must always be worn while in a chemistry laboratory. This is not simply due to common sense, but also government regulations. Your eyes are the only pair you have! Even if you are not performing an experiment, most laboratories are equipped with working equipment and/or chemicals requiring eye protection at all times. Eye protection must contain shatter-proof lenses with side shields. Safety goggles or glasses with side shields are common.

If a chemical gets into your eyes, you must quickly get to the eyewash and flush your eyes with water, while rolling them back and forth, for no less than 15 minutes. Eye contacts should **not** be worn in the laboratory; however, if you have eye contacts in, you must try to remove them immediately. Your instructor will help you locate the eyewash **prior to performing any experiments.**

Proper Attire

Clothing that covers your body and fits comfortably should be worn. Clothing that is loose can be a hazard and knock over glassware or catch fire in a flame. Clothing that is tight will hold chemicals close against your skin. Shorts, skirts, or skorts should not be worn. Shirts should cover your stomach and shoulders (preferably your arms too). Shoes should protect your feet and not be made of cloth (chemicals can leak through) or open-toed. Leather shoes are preferred. Jewelry should be removed; watches and rings can hold chemicals against the skin. Hair should be pulled back and eye contacts removed.

If corrosive chemicals come in contact with the hands, they should be rinsed immediately in the sink and your instructor notified. If corrosive chemicals are spilled on your clothing, all of your clothing must be removed and you must quickly get to the safety shower. The instructor should be notified and will evacuate the laboratory. Although embarrassing, it is important that all your clothing be removed such that the chemical can be quickly flushed from the skin. Wet clothing has a tendency to hold chemicals against the skin while being washed.

Fire Safety

Occasionally, fires do occur in the laboratory. Small fires can be extinguished by covering them with a container, such as a beaker, to eliminate oxygen. Moderately sized fires require the use of a fire extinguisher. Inform your instructor of these fires, as there are many different types of fire extinguishers. Large fires require the evacuation of the building and notifying the fire department. Your instructor will use discretion.

If your clothing catches fire, the best extinguisher is water from the safety shower. If a safety shower is not available, you can fall to the floor and cover yourself with a fire blanket. Make sure you are aware of the location of the safety shower and fire blanket **prior to performing any experiments.** Synthetic clothing fibers melt during combustion and produce hot liquids that stick to the skin. It is best to wear natural fibers, such as cotton, for fire safety when working in the laboratory.

Volatile and Toxic Chemicals

Many volatile and toxic chemicals must be used in a fume hood. Many toxic fumes have a pungent smell that can be overpowering (e.g., ammonia). Other toxic fumes may not be as noxious, but may cause respiratory or other health damage. It is important not to breathe either of these fumes. Fume hoods are enclosed workspaces that draw these fumes away from the user and out of the building. Most fume hoods have a safety-glass window that can be used to shield the user from violent or vigorous reactions. Fume hoods can also be used to remove flammable fumes that may pose a fire hazard.

Because potentially hazardous chemicals are found in the laboratory, it is important not to eat, smoke, or put on make-up while in the laboratory. All of these actions may potentially lead to the inadvertent ingestion of a hazardous chemical; in addition, smoking is a fire hazard.

Cuts, Burns and Spills

The most common accidents in a chemistry laboratory are cuts, burns, and spills. All spills must be reported to the instructor at the time of occurrence. Spilled chemicals can damage floors and benches; but more importantly, a student may inadvertently step into, slip on, or lean against a hazardous chemical. Spills must be cleaned up for your safety and the safety of others.

Cuts typically occur due to broken glassware rather than knives. Caution must be exercised when using glassware. Excessive force while performing a procedure (e.g., inserting a glass tube into a rubber stopper) may cause glassware to break in your hands. Proper procedures must be followed. All broken glassware must be reported and cleaned up.

Burns typically occur due to the handling of hot glassware. Be sure to allow items to cool before handling with your bare hands, or use a glove. It is important to be aware of the location of the first aid kit for minor cuts and burns **prior to performing any experiments.**

Conclusion

Always be aware of all potential hazards involved in the procedures you are performing and the chemicals you are using. Be cautious and use common sense. Material Safety Data Sheets describing the potential hazards of most chemicals are available. Follow all safety precautions given to you by your instructor and those suggested in this manual. The laboratory experiments contained in this manual may be hazardous if materials are not handled properly or procedures are not followed correctly and therefore should be conducted only under the supervision of your instructor. The use of chemicals and lab equipment always require the utilization of proper safety precautions.

Experiment 1
Measurements

Introduction

The ability to accurately measure *amount* is fundamental to forensic science. Drug laws are often written specifying penalties for possession of a specific drug quantity; thus, the mass of certain drug evidence must be accurately determined. Mass refers to the amount of matter present. It is commonly recorded in units of grams. One gram is approximately the mass of one 3×5 note card. It is crucial that the forensic scientist become familiar with laboratory techniques and instruments capable of measuring mass.

The instrument commonly used to measure mass is the balance. Various types of balances are used by the forensic scientist differing in their appearance, operation, and level of precision permitted in a mass determination. In addition to measuring the mass of drug evidence, a balance is commonly used in a forensic laboratory to measure the mass of test reagents, compare bullets and cartridge casings, prepare quantities of evidence for analysis, and compare components of soil samples.

Equal in importance to mass determination is the ability to measure length and volume. Length measurements at the crime scene are essential when crime scene reconstruction requires the exact distance of items of evidence with respect to each. One way to determine the relative distance of items of evidence from each other within a room at a crime scene, without measuring all possible combinations, is to simply measure the distance of each item from two common walls. With these measurements, and the Pythagorean Theorem, relative distances can be determined.

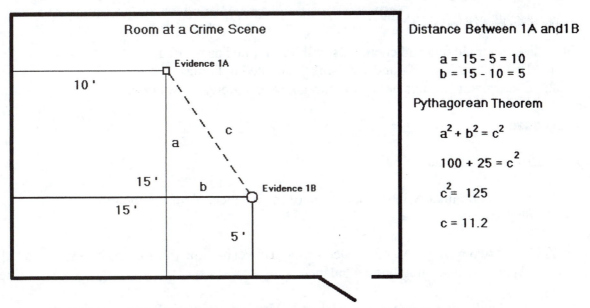

Figure 1.1 The use of the Pythagorean Theorem to determine distance.

Volume can be thought of as a three-dimensional measurement of length. Length is commonly measured using a ruler. For a regularly shaped solid (e.g., cylinder, square, rectangle, etc.), appropriate length measurements can be made and equations employed for volume calculation. Volume may be expressed in units of cubic centimeters (cm^3) or milliliters (mL). One cubic centimeter is exactly equal to one milliliter, and one centimeter is slightly less than ½ inch. For irregularly shaped solids and liquids, special volumetric glassware can be used to measure volume. Volumetric glassware commonly used in forensic laboratories includes the graduated cylinder, buret, and pipette. Each has unique markings allowing for accurate volumetric measurements. The type of volumetric glassware used will depend on the measurement being made and the accuracy desired. The graduated cylinder will be introduced in this experiment. The buret and pipette will be introduced in the Polymer Identification I experiment.

Purpose

To introduce the student to mass and volumetric measurement tools and techniques. Each student will become familiar with the correct operation and reading of a balance and a graduated cylinder. Students will also gain an understanding of how to correctly record a measurement to the degree of accuracy offered by the measuring device.

Materials

Obtain the following materials:
1. Two unknown mass samples
2. Two beakers (large enough for unknown mass samples)
3. One rectangular and one cylindrical solid
4. Two irregularly shaped solids
5. Ruler or tape measure
6. 25-mL graduated cylinder

The following materials and chemicals will be supplied as needed:
1. 10-, 100-, 250-, 500- and 1000-mL graduated cylinders
2. Graduated cylinders prepared with a known amount of water

Procedures

Part A Mass Determination

1. Become familiar with the operation of the balance. Your instructor may demonstrate.

2. Label two beakers A and B. Determine and **record** the mass of each in grams by placing each separately on a balance.

3. Place unknown mass #1 into beaker A. Using the same balance as in step 2, measure and **record** the combined mass of the beaker and the unknown sample in grams. Be sure to record all of the numbers displayed on the balance (even the

zeros). The mass of the unknown sample can be found by subtracting the mass of beaker A from the combined mass obtained.*

> * Small errors are likely to result when absolute mass measurements are made, i.e., when samples are placed directly on a tared ("zeroed") balance and a mass reading is directly obtained. Errors are more likely to occur and to be greater in magnitude when the balance has not been calibrated correctly or has been misused. The frequency of such errors can be decreased using the difference method suggested above.

4. Repeat step 3 using beaker B with the same unknown.

5. Repeat steps 3 and 4 using your unknown mass #2 instead of unknown #1.

6. Using a different balance from what was used in step 2, measure and **record** the combined mass of beaker A and unknown mass #1. To find the mass of the sample, you will subtract the mass of beaker A obtained using the balance in step 2 from the mass obtained in this step.*

> * You may find the calculated mass of your first unknown sample to be different from the mass calculated in step 6. This step demonstrates the importance of using the same balance for all measurements in a given procedure.

Part B Volume of Rectangular and Cylindrical Solids and Your Laboratory Room

1. Using a ruler, measure and **record** the length, width, and height of a rectangular object in centimeters. Understand to what fractional unit you can record your length using the supplied ruler. (Typically, one more decimal point is read than what is recorded on the ruler.) Volume can be calculated using the following equation. Units will be in cubic centimeters or milliliters.

$$Volume = length \times width \times height \hspace{3cm} 1.1$$

2. Measure and **record** the length, width, and height of your laboratory room in meters. If your laboratory is not in the shape of a single rectangle, you may need to measure the length width and height of several rectangular regions and add them together. Consult your instructor.

3. Using a ruler, measure and **record** the height and diameter of the cylindrical object in centimeters. When measuring the diameter, make sure that your ruler crosses the center of the circle. Volume can be calculated using the following equation. Units will be in cubic centimeters or milliliters.

$$Volume = 3.14 \times \left(\frac{diameter}{2} \right)^2 \times height \hspace{3cm} 1.2$$

Part C The Graduated Cylinder and Volume of an Irregularly Shaped Solid

1. Several graduated cylinders of varying sizes filled with different amounts of liquid have been prepared. Determine and **record** the volume of each liquid. The accuracy in volume may be different for each graduated cylinder. Make sure that you understand how to read each volume (i.e., looking at the bottom of the meniscus) and to what fractional unit of volume each graduated cylinder may be recorded.*

    * Step 1 introduces you to the graduated cylinder. Since it may be your first time using a graduated cylinder, make sure that you ask questions. You may want to show your answers to your instructor in order to make certain that you know how to read a graduated cylinder before proceeding.

2. Fill your 25-mL graduated cylinder approximately half way with water. Determine and **record** the exact volume of water to the accuracy of the graduated cylinder.

3. Carefully introduce your small irregularly shaped solid into the graduated cylinder. Make certain that no water is lost, that the solid is completely submerged, and that no bubbles are present. Notice the rise in water level.

4. Determine and **record** the new volume. The volume of your unknown solid will be the difference between your initial and final volumetric readings.

5. Repeat steps 2–4 for the large irregularly shaped solid using a 1000-mL graduated cylinder.

Part D Relative Distances of Crime Scene Evidence and the Pythagorean Theorem

1. Place a graduated cylinder on your bench top. Assume this to be one item of evidence. Also place a beaker and a test tube at separate locations throughout the laboratory on bench tops.

2. Measure the distance of the graduated cylinder, beaker, and test tube to two determined walls in meters.

3. Clean up when finished.

Experiment 1 Worksheet

Results and Observations

Part A Mass Determination

Unknown #1

Mass of Beaker A		Mass of Beaker B	
Combined A Mass		Combined B Mass	
Mass of Sample		Mass of Sample	

Unknown #2

Mass of Beaker A		Mass of Beaker B	
Combined A Mass		Combined B Mass	
Mass of Sample		Mass of Sample	

Unknown #1 (different balance)

Mass of Beaker A (step 2)	
Combined A Mass (step 6)	
Mass of Sample	

Part B Volume of Rectangular and Cylindrical Solids and Your Laboratory Room

Rectangular Solid

Width	Length	Height	Volume, cm^3 or mL

Laboratory Room

Width	Length	Height	Volume, m^3

Cylindrical Solid

Diameter	Height	Volume, cm^3 or mL

Part C The Graduated Cylinder and Volume of an Irregularly Shaped Solid

Graduated Cylinder Water Levels

Graduated Cylinder	10 mL	25 mL	100 mL	500 mL	1000 mL
Volume, mL					

Volume of Small and Large Irregularly Shaped Solids

	Small Solid	Large Solid
Initial Volume, mL		
Final Volume, mL		
Solid Volume, mL		

Part D Relative Distances of Crime Scene Evidence and the Pythagorean Theorem

Distance of Graduated Cylinder, Beaker, and Test Tube from the Walls

	Wall #1	Wall #2
Graduate Cylinder		
Beaker		
Test Tube		

Experiment 1 Questions

1. Why is it important to always use the same balance during the course of an experiment? You may discuss your results to support your answer.

2. Often, errors result when absolute mass determinations are made. For this reason, the difference method was used, requiring the sample to be placed in a container. Containers are commonly used for other reasons when measuring mass. What might some of these reasons be?

3. Mass is different from weight. Weight is a force that measures the pull of gravity on an object of a given mass and is measured in pounds or Newtons. The pull of gravity slightly changes as one increases or decreases in elevation. Also, the pull of gravity on the Earth is much greater than the pull of gravity on the Moon. Why is it important that a scientist measure mass and not weight?

4. Use the Pythagorean Theorem to determine the distances of the beaker from the test tube, the beaker from the graduated cylinder, and the graduated cylinder from the test tube. Show your work.

Experiment 2
Polymer Identification I

Introduction

Polymers are very large, long-chain molecules composed of repeating units called monomers. The monomer, or combination thereof, typically identifies the polymer. There are many natural polymers, such as DNA, proteins, and cellulose. Plastics, however, are synthetic polymers.

Plastics may be classified by one of seven recycling codes. Recycling code 1 identifies plastics composed of polyethylene terephthalate (PETE). PETE is commonly used for the production of soft drink bottles and polyester clothing fibers. Recycling code 3 identifies polyvinyl chloride (PVC) plastic products. PVC is used for some shampoo containers and spray bottles, pipes, and a not-so-common textile fiber called vinyon. Recycling codes 2 and 4 indicate high-density (HDPE) and low-density polyethylene (LDPE) plastics, respectively. HDPE is used for strong, rigid containers such as buckets, while LDPE is commonly used for shopping bags and some container lids. Code 5, polypropylene (PP), is used for household storage containers and Ziplock bags. Both polypropylene and polyethylene are used for the production of olefin textile fibers. Code 6 is polystyrene. It is most commonly used for the production is Styrofoam® cups, containers, and insulators. Code 7 identifies all other plastics.

In addition to a difference in physical appearance and usage, most plastics also differ in their density. Density is a physical property defined as the mass of a substance per unit volume. It is commonly expressed in units of grams per milliliter (i.e., g/mL).

$$Density = \frac{mass}{volume} \qquad\qquad 2.1$$

The mass of a substance is often measured using a balance, while the volume may be measured using a ruler or suitable volumetric glassware (as was done in Experiment 1). If volume stays constant and mass increases, density also increases; however, if the volume of a substance increases with little or no change to its mass, density decreases (e.g., as water freezes it expands; or its volume increases while its mass stays constant, making ice less dense than water).

It must be understood that most techniques used to measure density exploit average sample density. Small density variations potentially found within a substance are not considered, and the calculated density becomes the average density of the whole substance. For example, small air bubbles trapped in or on an object having a lesser density than the object will decrease the measured density of the object. Only when a substance is homogenous (same throughout), is the calculated density the true uniform density of the substance. Thus, contaminants found within or on a sample can significantly alter the sample's determined density.

Density serves as a very powerful comparative forensic tool. A density comparison analysis is routinely performed when attempting to compare an unknown sample with a known sample for the purpose of determining a common origin. The relative density of a solid sample (e.g., glass or plastic) can be determined by simply placing the sample into a solution of known density. If the sample has a density greater than that of the solution, it will sink; if it is less than that of the solution, it will float; and if it equals that of the solution, it will appear suspended in the solution. Density analyses are regularly performed on other items of evidence such as soil, plastic, glass, and wood. Density gradients can also be employed to visually compare the density of many items within a single sample. A density gradient is used in Experiment 3 to compare soil samples.

The purpose of this experiment is to distinguish plastic polymer samples. Plastics not only differ in their densities, but also in their burn characteristics in an open flame. For example, the burning of PVC may result in the production of a green-tinted flame due to the presence of chlorine. Many plastics melt and drip while burning, producing white smoke (e.g., PP, LDPE, and HDPE), while others produce black smoke and do not drip (e.g., polystyrene and PVC). This test is destructive and may only be used with an excessive amount of evidence.

Purpose

To introduce the student to density and flame analysis techniques for the distinguishing of polymer plastics. Each student will become familiar with the correct operation and reading of a pipette and buret.

Materials

Obtain the following materials:
1. Six standard plastic polymer samples (PETE, HDPE, PVC, LDPE, PP, and PS)
2. One unknown plastic sample
3. 100-mL graduated cylinder
4. Six 125-mL conical flasks (large test tubes or vials may be substituted)
5. Stoppers
6. 50-mL beaker
7. 250-mL beaker
8. Forceps

The following materials and chemicals will be supplied as needed:
1. Ethanol
2. Buret
3. 20-mL pipette
4. Calcium chloride
5. Distilled water
6. Bunsen burner

Procedures

Part A Preparation and Density Determination of Ethanol/H_2O Solutions (The Pipette)

1. Dispense 80 mL of ethanol into a 100-mL graduated cylinder. Fill the rest of the graduated cylinder with distilled water to the 100-mL mark. Pour the solution into a 125-mL conical flask and stopper to prevent evaporation.

2. Prepare three more solutions with relative amounts shown below in a similar manner, placing each new solution into a separate 125-mL conical flask.

   a. 55 mL of ethanol with distilled water to 100-mL mark
   b. 45 mL of ethanol with distilled water to 100-mL mark
   c. 100 mL of distilled water

3. Pipettes are volumetric glassware that can be used to dispense a desired volume of liquid with a great degree of accuracy and precision. The instructor will demonstrate how to use a pipette. Before proceeding, you may want to practice filling the pipette and dispensing its volume using distilled water.

4. Rinse a 20-mL pipette by filling it approximately ¼ full with the liquid to be measured. Swirl the liquid around with your finger (or pipette bulb) over the top opening, being careful not to have any liquid spill out the top. Dispense it into a waste beaker. Repeat two more times.

5. Measure and **record** in grams the mass of a 50-mL beaker. Make sure the beaker is clean and dry.

6. Fill the 20-mL pipette to mark with the first liquid prepared in step 1 and dispense it into the above beaker. Re-stopper the conical flask to prevent evaporation.

7. Measure and **record** in grams the combined mass of the beaker and liquid.

8. Calculate density.

9. Repeat steps 4–8 for the three other solutions prepared in step 2.

Part B Preparation and Density Determination of $CaCl_2$ Solutions (The Buret)

1. Measure 25.0 g of calcium chloride in a 250-mL beaker. Slowly add approximately 40–50 mL of distilled water so as not to exceed the 100-mL mark on the beaker. Swirl the beaker for several minutes until the calcium chloride has completely dissolved.*

* The dissolution of calcium chloride into water is extremely exothermic (heat-producing). The beaker will get very hot and may produce steam! It is advisable to hold the beaker above the solution level or use gloves.

2. Pour the calcium chloride solution into a 100-mL graduated cylinder. Rinse out the beaker with a small amount of distilled water and pour the rinse into the 100-mL graduated cylinder until the solution level reaches the 100-mL mark.

3. Next, pour the solution into a 125-mL conical flask and stopper to prevent evaporation.

4. Prepare one more calcium chloride solution in a similar manner using 50.0 g of calcium chloride.

5. Burets, similar to pipettes, are used to dispense a desired volume of liquid with great accuracy and precision. However, their application is quite different. Become familiar with how to read a buret. Your instructor will demonstrate.

6. Rinse the buret with several milliliters of the first solution prepared in steps 1 and 2 of this section. You do not need to fill the entire buret. Partially fill the buret and turn it on its axis while pouring the solution out so as to rinse the inside walls.

7. Fill the buret to just below the 25.00-mL mark with the first solution. Re-stopper the 125-mL conical flask to prevent evaporation. Open the stopcock and allow 3–4 mL to flush the tip into an empty beaker and remove any bubbles from the tip.

8. Measure and **record** in grams the mass of a 50-mL beaker. Make sure the beaker is clean and dry.

9. Determine and **record** the initial buret reading in milliliters. Into the above beaker, dispense approximately 15 mL of the liquid solution. Determine and **record** the final buret reading. Recognize that the markings on a buret are backward from that of a graduated cylinder. The amount of liquid dispensed will be your final buret reading minus your initial buret reading.

10. Measure and **record** in grams the combined mass of the beaker and liquid.

11. Calculate density.

12. Repeat steps 6–11 for the second calcium chloride solution.

Part C Density of Polymers with Respect to the Above Liquids

1. In the above two sections you prepared and determined the density of liquid solutions that will allow you to distinguish six different plastic polymers (PET, HDPE, PVC, LDPE, PP and PS).

2. Obtain six plastic polymer standards. Un-stopper the first solution prepared in Part A and add each standard plastic polymer. **Record** whether the samples float or sink.* If the sample floats, its density is less than that of the solution. If it sinks, its density is greater.

   * Many samples may appear to float due to surface tension. Gently depress any samples that appear to be floating, completely submerging them. If they rise back to the surface, they are truly floating samples. Also, many samples may acquire bubbles on their surface. It is important to remove these bubbles with a glass stirring rod or spatula.

3. Remove each sample placed into the first solution and dry thoroughly.

4. Continue adding and removing the plastic polymer standards from all of the solutions prepared in parts A and B. **Record** your observations. You should begin to be able to distinguish the standards.

Part D Polymer Flame Tests

1. Obtain and light a Bunsen burner in a fume hood. Using forceps, place a small amount of each plastic polymer standard into the flame. **Record** the appearance of the flame and smoke. **Record** whether or not the standard dripped while burning.

2. Repeat step 1 for all plastic polymer standards.

Part E Identifying an Unknown Polymer Sample

1. Obtain an unknown plastic polymer sample. **Record** its unknown number. Determine in which prepared solutions it floats and sinks.

2. Conduct a flame test on the sample and **record** your observations.

3. Clean up when finished. Dispose of all solutions appropriately.

Name:_____   Date:_____

Experiment 2 Worksheet

Results and Observations

Part A Preparation and Density Determination of Ethanol/H_2O Solutions (The Pipette)

Solution 1 (80 mL ethanol)

---	Beaker	Beaker + Sample	Sample
Mass, g			

Volume of pipette, mL:_____

Density, g/mL:_____

Solution 2 (55 mL ethanol)

---	Beaker	Beaker + Sample	Sample
Mass, g			

Volume of pipette, mL:_____

Density, g/mL:_____

Solution 3 (45 mL ethanol)

---	Beaker	Beaker + Sample	Sample
Mass, g			

Volume of pipette, mL:_____

Density, g/mL:_____

Solution 4 (distilled water)

---	Beaker	Beaker + Sample	Sample
Mass, g			

Volume of pipette, mL:_____

Density, g/mL:_____

Part B Preparation and Density Determination of CaCl₂ Solutions (The Buret)

Solution 5 (25 g calcium chloride)

---	Beaker	Beaker + Sample	Sample
Mass, g			

---	Initial	Final	Sample
Volume, mL			

Density, g/mL:_____

Solution 6 (50 g calcium chloride)

---	Beaker	Beaker + Sample	Sample
Mass, g			

---	Initial	Final	Sample
Volume, mL			

Density, g/mL:_____

Part C Density of Polymers with Respect to the Above Liquids

	PETE	HDPE	PVC	LDPE	PP	PS
Solution 1						
Solution 2						
Solution 3						
Solution 4						
Solution 5						
Solution 6						

Part D Polymer Flame Tests

	PETE	HDPE	PVC	LDPE	PP	PS
Flame						
Smoke						
Drips						

Part E Identifying an Unknown Polymer Sample

Unk #_____	Unknown Plastic Polymer Sample
Solution 1	
Solution 2	
Solution 3	
Solution 4	
Solution 5	
Solution 6	
Flame	
Smoke	
Drips	

Experiment 2 Questions

1. Suppose that a balance consistently gave higher readings than true values. How would this affect the calculated density of the solutions in parts A and B?

2. Knowing that air is less dense than plastic, what would happen to the relative density of a plastic sample with respect to the solution, if an excessive number of air bubbles were allowed to accumulate on the plastic surface? Explain.

3. If the density of a liquid is 0.789 g/mL, what would the volume of 1569.7 grams of the liquid be?

4. Based on the determined densities of the six prepared solutions and the information obtained in part C, determine the range in density of each plastic polymer standard.

5. Assume that the plastic polymer unknown sample was obtained at a crime scene after a hit-and-run. It is suspected that the sample came from a container crushed during the commission of the crime. A suspect has been located, and the bottom portion of a crushed container containing drugs was found in his possession. It is surmised that the victim (being the original owner of the container and drugs) was fleeing on foot with the drugs in the container from the suspect. The suspect hit the victim with his car in an attempt to obtain the drugs. In doing so, he ran over the container. In an attempt to recover the drugs, the suspect placed what he could find into the bottom portion of the crushed container and left. The suspect claims that the plastic fragments found at the crime scene did not come from his container. His container is made of HDPE. Determine the range in density of the unknown plastic polymer. Based on your density and flame test results, could the unknown sample have originated from the suspect's container? If not, what is the identity of the unknown plastic?

Experiment 3
Soil Examination

Introduction

To a forensic scientist, soil is any disintegrated material (natural or artificial) that lies on or near the Earth's surface. Disintegrated material is not limited to dirt, but may also include brick, cement, safe insulation, sawdust, etc. Oftentimes, a soil analysis is required during an investigation. Soil is typically class evidence. This means that rarely can a soil sample associated with a suspect be linked exclusively to soil found at a crime scene. However, as with all class evidence, it may serve to corroborate an alibi, aid in crime scene reconstruction, and/or exonerate a suspect. Soil found on the clothes and/or shoes of a suspect may serve to link him/her to a crime scene. Soil evidence should not be overlooked even if the crime scene location is not known. Occasionally, the composition of a soil sample may be unique enough to a particular area and aid in locating the crime scene.

Purpose

To introduce the student to techniques commonly used for soil comparison analyses. Students will become familiar with visual, pH, color, density, and particle-size analysis techniques.

Materials

Obtain the following materials:
1. Two soil samples
2. Two test tubes
3. Two 25-mL graduated cylinders
4. Seven 1-dram vials
5. Test tube rack

The following materials and chemicals will be supplied as needed:
1. Litmus paper
2. Plastic disposable pipettes
3. Spot plate
4. Methanol and bromoform (or calcium chloride)
5. Mechanical or manual sieves

Procedures

Part A Visual Characteristics

1. Pour out a small amount of soil samples A and B onto two separate pieces of white paper (use your own). Carefully exam each sample for its physical

characteristics. **Record** both similar and unique characteristics of each. Do not hurry through this step. Take your time and be meticulous.

Part B pH Determination

1. Place a small quantity of soil sample into a test tube. Add 1 to 2 mL of distilled water. Cap the test tube and shake vigorously for 30 seconds. Place the test tube in a test tube rack and allow the soil to settle.

2. With a disposable plastic pipette, draw out a small quantity of water and apply to red litmus paper.* Determine and **record** the pH by comparing the color developed to the color chart on the litmus paper container. Repeat with blue litmus paper.

    * Litmus paper changes color dependent on solution pH. The color change can be used to determine the approximate pH of the soil sample.

3. Repeat steps 1 and 2 for the second soil sample using similar soil and water amounts.

Part C Color Determination

1. Place a small amount of soil into two side-by-side wells of a spot plate. Just below or above the two wells, place a small amount of the second soil sample into two side-by-side wells of the same spot plate.

2. Using a plastic pipette, add a few drops of distilled water to one of the wells of each soil sample. Mix well using the tip of the pipette. Compare and **record** the similarities and differences in color of the wet and dry soil samples.

Part D Density Gradient

1. In a fume hood, using plastic disposable (graduated) pipettes, prepare the following seven solutions in separate 1-dram vials.*
    a. 3.0 mL pure bromoform (density 2.89 g/mL)
    b. 0.5 mL methanol, 2.5 mL bromoform
    c. 1.0 mL methanol, 2.0 mL bromoform
    d. 1.5 mL methanol, 1.5 mL bromoform
    e. 2.0 mL methanol, 1.0 mL bromoform
    f. 2.5 mL methanol, 0.5 mL bromoform
    g. 3.0 mL pure methanol (density 0.88 g/mL)

    * Methanol is a volatile and very flammable organic compound. Bromoform can irritate the skin, eyes, nose, throat, and lungs. At high exposure, it may cause headaches and dizziness. It is very important that both of these chemicals be used in the fume hood.

2. Carefully pour each of the solutions prepared in step 1, beginning with pure bromoform, into one 25-mL graduated cylinder.* Cap the top with a small cork stopper to prevent evaporation.

   * You are preparing a density gradient, so it is very important that the solutions do not mix. Pour each successive solution slowly down the inside glass wall of the cylinder.

3. Repeat steps 1 and 2, preparing a second density gradient in a second 25-mL graduated cylinder.

4. Place a small amount (~0.1 g) of each soil sample into each of the graduated cylinders. Most particles will reach their suspended depth within a few minutes.* After 10 minutes, compare and **record** the density gradient profiles of the two samples.

   * Preferably, the samples would sit overnight, allowing for minor adjustments in density and grain position.

Part E Particle Size Distribution

1. Disassemble the sieves of a mechanical size separator. Measure and **record** the mass of each sieve before the soil has been added (initial mass).

2. Restack the sieves, making sure that the sieve with the largest holes is on top and the sieve with the smallest holes on the bottom.

3. Measure and **record** the mass of approximately 1–2 tablespoons of soil and add to the top sieve and mechanically shake for approximately 5 minutes.*

   * If the stack of sieves do not contain a container to catch the finest particles, you will need to measure the mass of a piece of paper and shake the stack of sieves over the paper.

4. Carefully disassemble the sieves so as not to lose any soil. Measure the mass of each sieve containing the soil (final mass). The sample mass will be the difference between the initial and final sieve masses. Percent sample will be the sample mass of the soil in a particular sieve divided by the total sample mass multiplied by 100%.

5. Repeat for the second soil sample.

6. Clean up when finished.

Name:_____   Date:_____

Experiment 3 Worksheet

Results and Observations

Part A Visual Characteristics

	Differences	Similarities
1		
2		
3		
4		
5		
6		
7		

Part B pH Determination

Red Litmus Color A		Red Litmus Color B	
Blue Litmus Color A		Blue Litmus Color B	
Estimated pH		Estimated pH	

Part C Color Determination

Dry Color A		Dry Color B	
Wet Color A		Wet Color B	

Part D Density Gradient

o Draw a picture for the distribution of soil particles in each density gradient. Discuss their similarities and differences.

Part E Particle Size Distribution

Total Mass of Sample A

Sample A	Initial Mass	Final Mass	Sample Mass	Percent Sample
Sieve #1 (top)				
Sieve #2				
Sieve #3				
Sieve #4				
Sieve #5				

Total Mass of Sample B

Sample B	Initial Mass	Final Mass	Sample Mass	Percent Sample
Sieve #1 (top)				
Sieve #2				
Sieve #3				
Sieve #4				
Sieve #5				

Experiment 3 Questions

1. Based on your results, do the two samples compare? Make sure you discuss the results of each test performed.

2. Is it possible to have two soil samples differ in all tests performed in this experiment and still have a common origin? Explain.

3. Is it possible to have two soil samples compare in all tests performed in this experiment and **not** share a common origin? Explain.

4. What would happen to your results if all of the solutions prepared for the density gradient were mixed uniformly with each other upon pouring each into the 25-mL graduated cylinder?

Experiment 4
Thin Layer Chromatography Analysis

Introduction

Thin layer chromatography (TLC) is a technique employed to separate compounds found within a mixture. Compared with other separation techniques (e.g., gas chromatography, liquid chromatography, and capillary electrophoresis), TLC is relatively simple and requires only a minimal amount of equipment. The other mentioned techniques require expensive instrumentation. Because of its simple nature, TLC is frequently used in forensic science to compare compounds of a sample with a known origin to compounds of a sample with an unknown origin. For example, the compounds comprising a fiber or ink dye may be separated and compared to a fiber or ink dye found at a crime scene. The separation not only generates a profile representative of the mixture, but may also be used for presumptive compound identification. This is done by noting the position of the separated compounds in the separation profile. Forensic scientists commonly employ this technique for presumptive drug identification. The mixture of compounds found in pen inks will be compared and analyzed in this experiment using TLC.

Chromatography (literally meaning color writing) requires both a stationary and a mobile phase for compound separation. Compounds to be separated are carried over the stationary phase by a moving mobile phase. In order to understand chromatography, it is best to first appreciate compound attractions. When the physical properties of two compounds are similar, they have an attraction (i.e., affinity) for each other and will easily and regularly associate with each other. For example, water and oil do not mix because their physical properties are quite distinct; however, water and isopropyl alcohol (rubbing alcohol) have similar properties and mix well in all proportions. We might say that isopropyl alcohol has a "high affinity" for water, while oil has a "low affinity" for water. This selective interaction of compounds with other compounds is the basis of chromatography. Because the mobile phase is moving and carries the compounds to be separated over an immobile stationary phase, the compounds with a higher affinity (attraction) for the stationary phase will not pass over the stationary phase as quickly as the compounds that have a higher affinity for the mobile phase. It is this selective interaction of the compounds, based on their differing physical properties, with both the stationary and mobile phases that allow compounds to be separated from each other. The quality of a separation can be measured by how well compounds of interest are separated.

The separation system containing the stationary phase can be an enclosed tube or capillary through which mobile phase passes. Or more simply, it may also be a plate containing stationary phase over which mobile phase passes. TLC employs a flat glass, paper, or plastic plate serving as either the stationary phase or the substrate to which a stationary phase is affixed. The compounds are separated on the plate as a mobile phase is allowed to travel over the plate in a vertical direction due to capillary action. The same phenomenon is seen when an M&M is left on a wet paper towel and the compounds comprising the dye are separated and seen as individual colors (a visual example of true "color writing"). The paper towel serves the role of the stationary phase and the water the role of the mobile phase. Separated compounds in TLC are often not very colorful and must be visualized using chemical developers or ultraviolet lighting. The relative position of each compound with respect to the distance traveled up the plate by the mobile phase after separation (known as retardation factor, R_f) becomes representative of the compound and can be used for presumptive identification.

Purpose

To introduce the student to thin layer chromatography and its application for comparing inks. Students will experience the use of two different solvent systems for separating compounds found in ink. Results will be compared using both visible and ultraviolet lighting.

Materials

Obtain the following materials:
1. Two TLC plates
2. One 25-mL graduated cylinder
3. Two 400-mL beakers
4. One 250-mL beaker or test tube rack
5. Four test tubes
6. One watch glass
7. Test tube rack

The following materials and chemicals will be supplied as needed:
1. Isopropyl alcohol
2. Methanol
3. Scissors
4. Pencil
5. Capillary tubes (TLC spotters)
6. Ultraviolet light source

Procedures

Part A Prepare TLC Samples

1. Three known black markers were used to prepare three standards by writing the letters A, B, or C with each. Obtain these standards. One of the three markers, or a combination of markers, was/were used to write a number. Obtain one unknown and **record** its number. **Record** the appearance of all four samples.

2. Cut the paper containing the ink as small as possible, removing paper not containing ink. Crumple each into a little ball and place each into separate test tubes (i.e., A, B, C, and #).

3. Add 5–10 drops of methanol into each test tube containing the samples. Gently swirl each test tube so as to extract the ink from the paper. **Record** the appearance of the methanol for each sample.

4. Place all four samples upright in a test tube rack.

Part B Prepare TLC Plates and Separate Samples

1. In the hood, carefully pour 15 mL of isopropyl alcohol into a 25-mL graduated cylinder. Into the same graduated cylinder, carefully pour 5 mL of methanol, filling the graduated cylinder to the 20-mL mark.*

    * Both isopropyl alcohol and methanol are volatile and flammable organic solvents.

2. Pour the solution prepared above into a 400-mL beaker and cover the beaker with a watch glass.

3. Obtain a TLC plate. Using a pencil, draw a light line across the TLC plate ¼ – ½ inches from the bottom. Across the top of the plate, write A, B, C, and # (your unknown number) equally spaced as seen in Figure 4.1.

Figure 4.1. Example of TLC plate

4. Obtain a TLC spotter. Carefully break the narrow end of the TLC spotter to create a small opening if sealed. Partially fill the TLC spotter with solution A by tilting the test tube containing solution A and placing the narrow end of the spotter into the solution with your finger removed from the top opening of the spotter. Once the spotter has been partially filled, place your finger on the top opening of the spotter and remove it from the solution.

5. On the line marked close to the bottom of the TLC plate and inline with the letter A written at the top of the plate, carefully touch the spotter to the TLC plate, making a tiny spot of solution while still holding your finger over the top opening of the spotter. Wait a few seconds for the solvent to dry (or blow it dry) and repeat nine more times in the same spot. If the spotter will no longer deliver solution, away from your plate, briefly remove your finger from the top opening and then quickly replace it. If a small droplet appears on the bottom of the spotter, gently shake it off and continue spotting.

6. When finished, place the spotter either in the test tube containing the sample spotted or in a location where you may use it later.

7. Repeat steps 5 and 6 for samples B, C, and #.

8. Place the prepared TLC plate into the beaker containing the solvent prepared in step 1 with the side containing the sample spots down. If the solvent level in the beaker is higher than the line marked on the TLC plate, dump a little of the solvent into a waste beaker before inserting the TLC plate.*

   * It is important that when inserting the TLC plate into the beaker that the solvent does not cover the prepared spots. If this were to occur, sample would be lost to the solution before a separation could occur.

9. Cover the beaker with a watch glass to minimize evaporation and set it aside for 20–30 minutes.

Part C Prepare TLC Plates and Separate Samples Using a Different Solvent System

1. Using a 25-mL graduated cylinder, prepare 20 mL of mobile phase using isopropyl alcohol and methanol at a ratio different from that prepared in part B.* **Record** the ratio used.

   * The ability to separate compounds using TLC is a function of both the nature of the TLC plate (i.e., the stationary phase) and the mobile phase. It is difficult to change the nature of the TLC plate; however, we can very easily change the nature of the mobile phase by either changing the solvents that comprise the solution and/or changing their combining ratio.

2. Repeat steps 2–9 in part B using the newly prepared mobile phase.

Part D Examine Results

1. After approximately 20–30 minutes of separating time, remove each TLC plate from its beaker and immediately mark the height of the solvent front with a pencil. If you wait too long, the solvent will evaporate and you may not be able to

tell how far up the solvent moved. You will need to know how far the solvent traveled in order to answer one of the post-lab questions.

2. Examine the TLC plate for part B using regular lighting and ultraviolet lighting and determine your unknown. **Record** the appearance of the TLC plate under both regular ultraviolet lighting.

3. Compare the results obtained in part B with those obtained in part C. Record the appearance of the TLC plate for part C under ultraviolet lighting. Staple your TLC plates to your worksheet.

4. Clean up when finished.

Name:_____   Date:_____

Experiment 4 Worksheet

Results and Observations

Part A Prepare TLC Samples

Unknown #:_____

Appearance of Standards and Unknown Prior to Extraction

A	B	C	#

Appearance of Standards and Unknown Methanol Extracts

A	B	C	#

Part C Prepare TLC Plates and Separate Samples Using a Different Solvent System

- o   Solvent Ratio:_____

Part D Examine Results

- o Appearance of TLC Plate in Part B Using Ultraviolet and Regular Lighting (use words and drawings)

- o Appearance of TLC Plate in Part C Using Ultraviolet and Regular Lighting (use words and drawings) o Staple One TLC Plate Here and Label It (Your partner will staple the other.)

o Unknown #_____ contains (A, B, C, or B and C) _____.

Experiment 4 Questions

1. Discuss the differences in spot positions between results obtained for the TLC plates prepared in parts B and C. Offer an explanation as to why the differences might have occurred.

2. Sometimes, the following equation is used to presumptively identify a compound found in a mixture using TLC. R_f is the retardation factor, d is the distance the compound traveled, and D is the distance the mobile phase traveled. Calculate the R_f value (just for the part B separation) for one blue dot, one red dot, and the two prominent black/purple dots (for one standard) found in the middle of the TLC plate on all three standards.

$$R_f = \frac{d}{D}$$

3. Is it possible to have the two samples analyzed in this experiment compare and not share a common origin? Explain.

Experiment 5
Polymer Identification II (FTIR)

Introduction

The Fourier transform infrared spectrophotometer (FTIR) is one of the most commonly used instruments in the crime laboratory. It is used for drug, fiber, paint, plastic, and unknown substance analyses. The instrument is capable of generating a compound "fingerprint", i.e., a unique spectrum of infrared light absorbance, which can be used for compound identification. The wavelengths of absorbed infrared light, in conjunction with their relative intensities, serve as a means of individualization. All FTIR spectra are representative of compound structure, and all compound structures (except optical isomers) differ. Thus, FTIR spectra are very compound specific and can be matched to standard spectra stored in libraries for the purpose of compound identification.

The interaction of infrared light with matter, resulting in an FTIR spectrum, is specifically due to the absorbance of infrared light by the bonds of a molecule. The atoms comprising a chemical bond are constantly vibrating at a defined frequency (and energy). If the frequency of vibration of infrared light matches the frequency of vibration of atoms within a molecular bond, an absorbance will occur.

Just as a change in the mass of an object on a spring changes the frequency of vibration of the spring, a change in the mass of atoms found between a chemical bond changes the vibrational frequency of the bond. Typically, as the mass increases, the frequency decreases. In addition, the strength of the bond influences the vibrational frequency, i.e., as the bond strength increases (e.g., going from a single to a double bond) the frequency also increases. These structural relationships to infrared light absorbance can be used to link a peak found in an FTIR spectrum to a structural feature of the molecule (e.g., C=O, O-H, C-H, etc.).

Because the structures of most molecules differ, virtually all compounds have a unique FTIR spectrum. However, some compounds have similar spectra, while others have no spectra at all. The compounds with similar spectra may require a trained eye or an additional analysis for identification. Not all bonds within a molecule have a tendency to absorb infrared radiation even when the bond frequencies and energies match. These bonds are said to be "FTIR inactive." Some molecules are composed only of FTIR inactive bonds and do not have an FTIR spectrum (e.g., N_2). Despite these limitations, FTIR continues to be a very powerful forensic analysis tool.

Purpose

To introduce the student to the FTIR instrument and its generated spectra. Students will become familiar with the use of the instrument for the identification of plastic polymers. Students will also be able to link FTIR peaks with molecular structural features.

Materials

Obtain the following materials:
1. Six standard plastic polymer samples (PETE, HDPE, PVC, LDPE, PP, and PS)
2. One unknown plastic sample

The following materials and chemicals will be supplied as needed:
1. FTIR with attenuated total reflectance accessory

Procedures

Part A Acquisition of FTIR Spectra

1. Obtain one plastic polymer sample each of polyethylene terephthalate, high density polyethylene, polyvinyl chloride, low density polyethylene, polypropylene, polystyrene, and an unknown plastic polymer. **Record** their appearance.

2. Your instructor will demonstrate the proper use of the FTIR instrument. **Record** all procedures.

3. Obtain an FTIR spectrum of each of your samples including your unknown. Print out your results. Label each standard and unknown spectrum.

Part B Identification of Major FTIR Spectral Peaks

1. Using your textbook, identify at least two peaks in each of the standard and unknown spectra. Recognize that many peaks appear over a range of wavenumbers.

Part C Identification of an Unknown Polymer Sample

1. Using the obtained standard spectra, identify your unknown. It may or may not be one of the standard spectra (consult your instructor).

2. Your instructor may wish to demonstrate the identification of an unknown FTIR spectrum by performing a library search within the software of the instrument.

Name:_____   Date:_____

Experiment 5 Worksheet

Results and Observations

Part A Acquisition of FTIR Spectra

o   Record the appearance of each plastic polymer sample.

o   Briefly discuss the procedures for acquisition of FTIR spectra using your laboratory instrument.

Part B Identification of Major FTIR Spectral Peaks

o   Using your textbook, identify at least two peaks for each spectrum acquired in part A. Mark their identifications on your spectra. Turn in your spectra.

Part C Identification of an Unknown Polymer Sample

o   Visually compare your unknown spectrum to all six standard spectra. Which standard spectrum is most similar to your unknown spectrum? Explain.

o   Briefly discuss why your unknown spectrum does not match the other five standard spectra.

o Based on your above discussion, does the unknown spectrum match the most similar standard spectrum (i.e., can you identify it)?

Experiment 5 Questions

1. Briefly discuss what information is being acquired during an FTIR analysis. How can one sample be distinguished from another?

2. Often, wavenumbers (cm^{-1}) are used in FTIR analysis rather then frequency or wavelength. Convert 1234 cm^{-1} to both a frequency and a wavelength in nanometers.

3. Explain why there is no real visual difference in the FTIR spectra of the high-density and low-density polyethylene standards.

Experiment 6
Urine and Blood Analysis

Introduction

The presence of chemical species within urine and blood, with their respective concentrations, has routinely been used as a diagnostic tool for both illness and death. Nurses and doctors commonly request a blood and/or urine sample in an attempt to diagnose health conditions, while toxicologists regularly analyze blood and urine to aid medical examiners in the determination of cause of death.

The composition of both blood and urine changes as a function of health. Urine is often analyzed for the presence of proteins, glucose, and/or ketones. These species are not present to any appreciable degree in the urine of a healthy individual. Presence of one or more of these species at significant levels may indicate dehydration, kidney damage, or diabetes. In addition, urine and blood are both frequently analyzed for electrolyte levels of sodium and potassium, while blood is additionally analyzed for the presence of chloride, phosphate, calcium, and ammonium. All of these electrolytes are endogenous to both urine and blood and have defined normal levels. However, an increase or decrease in their normal levels may indicate a condition of heart failure, diabetes, or starvation.

Table 6.1 shows significantly high levels of electrolytes, proteins, glucose, and ketones (as indicated by an "x") in urine and blood of six potentially fatal medical conditions. Elevated levels for pregnancy are also shown. You will use Table 6.1, in conjunction with your results from this experiment of both simulated blood and urine samples taken from the assumed corpse of an unidentified female, to presumptively determine cause of death. In addition, you will determine whether or not she may have been pregnant.

Table 6.1 Elevated Levels in Urine (U) and Blood (B) as a Function of Medical Conditions

	Protien (U)	Ketones (U)	Glucose (U)	Sodium (U)	Potassium (U)	Sodium (B)	Potassium (B)	Chloride (B)	Phosphate (B)	Calcium (B)	Ammonium (B)
Dehydration		x		x		x	x	x			
Starvation		x		x							
Heavy Metal Poisoning	x										
Kidney Failure	x		x		x		x	x	x		
Diabetes-related	x	x	x	x			x				
Heart Failure	x										x
Pregnancy-related			x	x							

Purpose

To introduce the student to blood and urine analysis techniques of common electrolytes, proteins, glucose, and ketones. Students will become familiar with common precipitating and color-changing reactions.

Materials

Obtain the following materials:
1. Standard simulated blood and urine samples of potassium and sodium

2. Standard simulated blood samples of chloride, calcium, ammonium, and phosphate
3. Blank simulated urine and blood samples
4. 10 test tubes
5. Test tube clamp
6. Bunsen burner
7. Matches or striker
8. Hot plate or ring stand with ring and wire mesh to boil water
9. Thermometer
10. 10 graduated 1-mL disposable pipettes
11. 250-mL beaker

The following materials and chemicals will be supplied as needed:
1. Nichrome wire
2. Unknown simulated blood sample
3. Unknown simulated urine sample
4. 3 M nitric acid
5. 0.05 M silver nitrate
6. 0.1 M ammonium oxalate solution
7. 6 M sodium hydroxide
8. Red litmus paper
9. 0.02 M Ammonium molybdate solution
10. Glacial acetic acid
11. 0.1 M Sodium nitroprusside solution
12. Concentrated ammonium hydroxide
13. Benedict's reagent

Procedures

Part A Qualitative Analysis of Blood and Urine Electrolytes

1. Obtain standard simulated blood and urine samples of potassium and sodium. Obtain standard simulated blood samples of chloride, calcium, ammonium, and phosphate. Also obtain one simulated blood and one simulated urine sample known not to contain any of the above species (blank samples). Label all of your samples.

Test for Sodium and Potassium Ions:

2. Obtain approximately 8 inches of nichrome wire. Bend a small loop at one end to allow trapping of a small amount of liquid sample.

3. Prepare and light your Bunsen burner. Dip the loop of your nichrome wire into your blank simulated blood sample. Introduce the wire loop into the flame and **record** your results.*

* You will be performing a flame test. The color of the flame prior to the wire glowing orange will be representative of the species present (i.e., potassium or sodium). You may wish to **record** the appearance of the flame both prior to and after the wire begins to glow.

4. Repeat step 3 with your standard simulated blood sodium sample.

5. Repeat steps 3 and 4 with your blank and standard simulated urine sodium samples, respectively.

6. Repeat steps 3–5 using your standard simulated blood and urine potassium samples.

7. Turn off your Bunsen burner.

Test for Chloride Ion:

8. Using a graduated disposable pipette, dispense 1 mL of standard simulated blood chloride sample into a test tube and 1 mL of blank simulated blood into a separate test tube.

9. Carefully dispense 3–5 drops of silver nitrate solution into each test tube. **Record** both initial and subsequent observations for both test tubes.*

   * This test and all subsequent tests in this lab will result in the formation of a precipitate, or a color change. The precipitate formation or color change should not occur in the blank sample. Be sure you accurately describe your results such that you will be able to recognize them later when analyzing an unknown sample.

10. Dispose of each sample in an appropriate waste container. Rinse your test tubes clean.

Test for Phosphate Ion:

11. Fill a 250-mL beaker approximately half full with water. Place the beaker on a hot plate or ring stand with a Bunsen burner. Slowly begin heating the water to approximately 65–70°C (do not boil).

12. While the water is heating, dispense 1 mL of standard simulated blood phosphate sample into a test tube and 1 mL of blank simulated blood into a separate test tube.

13. Add 3–5 drops of 3 M nitric acid and 20 drops of ammonium molybdate solution to each test tube.

14. Place the test tubes upright in the warmed water. Allow them to sit for 5–10 minutes. **Record** your observations.

15. Turn off your hot plate or Bunsen burner. Dispose of your samples in an appropriate waste container and rinse your test tubes clean. You may need to use a test tube brush for cleaning.

Test for Calcium Ion:

16. Dispense 1 mL of standard simulated blood calcium sample into a test tube and 1 mL of blank simulated blood into a separate test tube.

17. Add 3–5 drops of ammonium oxalate solution to each test tube. **Record** your observations. Results may not be immediate. If you are having trouble seeing a difference between the samples, you may wish to add a drop of concentration ammonium hydroxide (consult instructor).

18. Dispose of your samples in an appropriate waste container and rinse your test tubes clean.

Test for Ammonium Ion:

19. Dispense 1 mL of standard simulated blood ammonium sample into a test tube and 1 mL of blank simulated blood into a separate test tube. Additionally, add approximately 0.5 mL of 6 M sodium hydroxide to each sample.

20. Wet two strips of red litmus paper with distilled water and drape one over each opening of your two test tubes. Place a test tube clamp near the top of the each test tube holding the litmus paper in place.

21. Prepare and light your Bunsen burner. Position the test tube containing the blood ammonium sample over the open flame of the Bunsen burner. Hold the test tube at a slight angle with the opening away from your fellow students (and instructor).* Shake the test tube generously to keep the solution mixing while in the flame. Avoid boiling the contents. Remove quickly from the flame if boiling occurs. After 20–30 seconds **record** the appearance of the red litmus paper.

    * Your sample may become superheated and the contents may be violently dispelled from the test tube.

22. Repeat steps 21 with your blank simulated blood sample test tube.

23. Turn off your Bunsen burner. Dispose of your samples in an appropriate waste container and rinse your test tubes clean.

Part B Qualitative Analysis of Urine Proteins, Glucose, and Ketones

1. Obtain standard simulated urine samples of protein, glucose, and ketones. Label your samples if this has not already been done.

Test for Proteins:

2. Dispense approximately 1–2 mL of standard simulated urine protein sample and 1–2 mL of blank simulated urine sample into two separate test tubes.

3. Prepare and light your Bunsen burner. Hold the urine protein test tube over the flame of the Bunsen burner. Follow all precautions given in part A when heating a test tube in an open flame. Remove the test tube from the flame just as it begins to boil. The precipitate formed in this experiment appears as "soapy" bubbles or froth on the inside of the test tube. **Record** your observations.

4. Repeat step 3 with your blank simulated urine sample.

5. Turn off your Bunsen burner. Dispose of your samples in an appropriate waste container and rinse your test tubes clean.

Test for Glucose:

6. Set up an apparatus similar to that in part A, step 11; however, the water will need to begin boiling for this procedure.

7. Add 5–10 drops of standard simulated urine glucose sample and 5–10 drops of blank simulated urine into two separate test tubes. Additionally, using a graduated disposable pipette, add 2 ml of Benedict's solution to each test tube.

8. After the water prepared in step 6 has begun to boil, place the test tubes into the boiling water and allow it to heat for 5 minutes. Remove from heat and allow it to cool. **Record** your observations.

9. Turn off your Bunsen burner or hot plate. Dispose of your samples in an appropriate waste container and rinse your test tubes clean.

Test for Ketones:

10. Dispense approximately 1–2 mL of standard simulated urine ketone sample and 1–2 mL of blank simulated urine sample into two separate test tubes. Additionally, add 1 mL of sodium nitroprusside solution to each.

11. In a fume hood, slowly add approximately 1 mL of concentrated ammonium hydroxide to each test tube, so as to form a separate layer on top of the solutions.

12. After 1–2 minutes, **record** your observations focusing on the interface between the two layers.

13. Dispose of your samples in an appropriate waste container and rinse your test tubes clean.

Part C Analysis of Unknown Blood and Urine Samples

1. Obtain an unknown simulated blood and urine sample. **Record** your unknown numbers.

2. Test your unknown simulated blood and urine samples with procedures in parts A and B. You do not need to run blank samples unless required by your instructor. **Record** all of your observations.

3. Determine in each case if the species being tested is present. Explain your reasoning.

4. Clean up when finished.

Name:_____   Date:_____

Experiment 6 Worksheet

Results and Observations

Part A Qualitative Analysis of Blood and Urine Electrolytes

- o   Observations

	Standard Sample	Blank Sample
Blood Sodium		
Urine Sodium		
Blood Potassium		
Urine Potassium		
Blood Chloride		
Blood Phosphate		

Blood Calcium		
Blood Ammonium		

Part B Qualitative Analysis of Urine Proteins, Glucose, and Ketones

o Observations

	Standard Sample	Blank Sample
Urine Proteins		
Urine Glucose		
Urine Ketones		

Part C Analysis of Unknown Blood and Urine Samples

o Observations   Unknown #s_____

	Unknown Sample	Present or Not Present? Explain.
Blood sodium		

Urine Sodium		
Blood Potassium		
Urine Potassium		
Blood Chloride		
Blood Phosphate		
Blood Calcium		
Blood Ammonium		
Urine Proteins		
Urine Glucose		
Urine Ketones		

Experiment 6 Questions

1. Table 6.1 shows significantly high levels of species tested in this experiment corresponding with potential health conditions. Assume that the tests you performed in this experiment only gave positive results (i.e., formation of color or precipitate) when the concentration of the species was greater than "normal" levels. Using Table 6.1 and your results for your unknown samples (taken from the deceased women), determine if the women may have potentially died from one of the listed conditions. Also determine if she could have been pregnant. Explain your reasoning.

2. Review your chemistry text and explain why the color of the flame is different for potassium and sodium standard samples. Hint: You will need to briefly discuss atomic structure, energy levels, and electrons.

3. Review you chemistry text and identify the precipitate products for the chloride and calcium tests.

Experiment 7
Blood Alcohol Concentration

Introduction

Throughout history, blood alcohol concentration has been determined in a variety of manners employing both chemical reactions and analytical instrumentation. Techniques have included the use of distillations, dichromate or permanganate salts, alcohol dehydrogenase enzymes, and ultraviolet/visible light spectrophotometry. A presently popular technique uses gas chromatography to directly measure alcohol concentration in whole blood samples.

Alternatively, breath alcohol concentration can be measured and related to blood alcohol concentration through an equilibrium volumetric ratio (i.e., 2100:1). The most common technique employs an infrared light spectrophotometer known as an Intoxilyzer. Suspects blow into the Intoxilyzer and the last portion of their breath is analyzed for the presence of alcohol. Most instruments can also identify common contaminants or false positives such as acetone and rubbing alcohol (i.e., isopropyl alcohol). The Breathalyzer is a more archaic instrument also used to measure breath alcohol concentrations. The analysis required the reaction of alcohol with potassium dichromate generating a color change. The extent of color change was subsequently measured using an instrument known as an ultraviolet/visible light spectrophotometer to determine alcohol content. A modified version of this technique will be used in this experiment to directly measure simulated blood alcohol concentrations.

Purpose

To introduce the student to the chemical reaction used in the Breathalyzer and infrared light absorbance in the Intoxilyzer to determine breath alcohol concentration. Each student will become familiar with the color changes used in the Breathalyzer and the interpretation of acetone and alcohol IR spectra.

Materials

Obtain the following materials:
1. Two 50-mL beakers
2. Five 1-dram vials
3. Spatula
4. One 25-mL graduated cylinder
5. Ten 1-mL disposable plastic pipettes
6. Glass stirring rod

The following materials and chemicals will be supplied as needed:
1. Acetone
2. Denatured ethanol
3. Potassium dichromate

4. Weigh boats

Procedures

Part A Determining Blood Alcohol Concentration Using Potassium Dichromate

1. Place a weigh boat onto the scale of an analytical balance. Tare the balance by pressing the "tare" or "zero" button.* The balance should read zero. Measure out approximately 0.05 g of potassium dichromate into the weigh boat. **Record** the exact sample mass.

    * A balance may be tared resulting in the deduction of the container mass prior to measuring sample mass. This technique will be used in this experiment rather than weighing by differences.

2. Using the same procedure used above, measure out approximately 0.005 g of silver nitrate in a separate weigh boat. **Record** the exact sample mass.

3. Empty the weighed potassium dichromate and silver nitrate into a 50-ml beaker. Dispose of the weigh boats.

4. Add 10 mL of distilled water to the above 50-ml beaker and mix by gently stirring or using a glass rod. **Record** the appearance of the solution.

5. In the hood, add 5 ml of concentrated sulfuric acid to the same beaker using a 1-mL plastic pipette.*

    * Sulfuric acid is very corrosive. If any gets on your hands, in your eyes, or on your clothes rinse thoroughly! Always inform the instructor of any accidents. It is best to rinse your hands after dispensing the sulfuric acid even if you don't think you have any on you.

6. Gently swirl the beaker (preferably in a fume hood) until all of the potassium dichromate has dissolved.* **Record** the appearance of the solution and set it aside where it will not spill.

    * The beaker will get hot as the water and sulfuric acid mix. Wait until the beaker is cool enough to carry before taking it to your workstation. Some of the silver nitrate may not dissolve completely.

7. Into a separate 50-ml beaker, add 20 mL of distilled water using a graduated cylinder. Using a 1-mL plastic pipette, add 20 drops of ethanol, holding the dropper at a 45-degree angle. This is your simulated blood alcohol solution.

8. Add the following to separate 1-dram vials:
    Vial #1: 1 mL of blood alcohol solution

Vial #2 ¾ mL of blood alcohol solution ¼ mL of distilled water
Vial #3: ½ mL of blood alcohol solution and ½ mL of distilled water
Vial #4: ¼ mL of blood alcohol solution and ¾ of distilled water
Vial #5: 1 mL of distilled water (this is your standard)
Vial #6: 1 mL of unknown blood alcohol concentration solution*

   * Acquire the unknown sample from the instructor. **Record** your unknown
     number.

9. In a fume hood, using a 1-mL plastic pipette, dispense 2 mL of prepared sulfuric
    acid solution into each of the above prepared 1-dram vials.*

   * The reaction begins when you add the sulfuric acid solution to the vial. You
     would like the reactions in all of the vials to begin as close in time to each other
     as possible. Do not delay in adding the 2 mL of prepared sulfuric acid solution
     to each successive 1-dram vial. Your partner may wish to fill three as you fill
     three. Use caution!

10. Place caps on the vials and shake all of them simultaneously. **Record** the initial
    color of all six vials.

11. Place the vials on your workbench in order and watch their color change. The
    color change will take several minutes. **Record** the appearance of all six vials at
    1-minute intervals for 15 minutes. During the color change process, there will be
    a time when the solution of all four of the vials will be distinguishable.* At this
    time, compare the color of the unknown to that of the four standard colors.
    Identify the standard vial that is most similar in color to that of the unknown.
    Record the milliliters of blood alcohol solution in your unknown.

   * It is important not to miss the time at which the colors are distinguishable. If
     you don't wait long enough, or if you wait too long, the colors will not be
     distinguishable.

12. Dispose of the solutions in the vials in a properly labeled waste container found in
    a fume hood.

13. Rinse the vials and their lids in tap water and then distilled water.

Part B Infrared Spectra

1. Acquire from the instructor infrared spectra for liquid ethanol and acetone. You
    will need these to answer question #3. Alternatively, acquire your own acetone
    and ethanol spectra using your laboratory FTIR.

2. Clean up when finished.

Name:_____   Date:_____

Experiment 7 Worksheet

Results and Observations

Part A Determining Blood Alcohol Concentration Using Potassium Dichromate

- Exact Mass of Potassium Dichromate:_____

- Exact Mass of Silver Nitrate:_____

- Appearance of Solution after Addition of Distilled Water

- Appearance of Solution after Addition of Sulfuric Acid

Unknown #: _____   Milliliters of Blood Alcohol Solution in Unknown.:_____

Appearance of Vials (Color)

	Vial #1	Vial #2	Vial #3	Vial #4	Vial #5	Vial #6 U
Initial						
1 min						
2 min						
3 min						
4 min						
5 min						
6 min						
7 min						
8 min						
9 min						
10 min						
11 min						
12 min						
13 min						
14 min						
15 min						

Experiment 7 Questions

1. If there are 45 drops of alcohol/mL, and the density of alcohol is 0.789 g/mL, what is the blood alcohol concentration of your unknown sample? (Hint: It will be higher than common blood alcohol levels. Also, blood alcohol concentration is defined in grams of alcohol/100 mL of blood.)

2. Potassium dichromate will react with more than just ethanol to generate a color change. Why would this be important information to know when using this technique to determine blood alcohol concentrations?

3. Some of the common wavelengths of light used in an infrared breath tester include 3.39, 3.48, and 9.50 micrometers (1 micrometer = 1×10^{-6} m). Convert the following numbers to cm (1 cm = 1×10^{-2} m) and then take the inverse of that number. List these values below and mark them on your IR spectra. Given these values and the acetone and ethanol spectra, briefly discuss how acetone could be identified as a contaminant.

Experiment 8
Arson and Accelerants

Introduction

In addition to using separation techniques to isolate desired compounds for the purpose of identification (e.g., drug identification), separation techniques can also be used to characterize a substance composed of a complex mixture of compounds (e.g., paint, accelerants, and dyes). The process of separating compounds within a mixture allows for the generation of a separation profile that is representative of the given mixture. The profile can then be used for mixture identification. Oftentimes, certain compounds characteristic of a given accelerant are individually identified employing a subsequent mass spectral analysis. Such additional information can further facilitate accelerant identification.

The above technique is routinely employed to identify the presence of liquid accelerant in arson debris using gas chromatography. Gasoline, as an example, is a common accelerant used in arson. It consists of a complex mixture of numerous aromatic and aliphatic organic compounds. The separation (using gas chromatography) and detection (using mass spectrometry) of these compounds allows for gasoline identification in arson debris. Even in a raging fire, many of these compounds are still found in charred absorbing material such as carpet, fabric, or wood due to the lack of oxygen deep in these materials. Accelerant identification is typically made by identifying a significant number of the compounds (but not necessarily all of them) comprising the accelerant.

Prior to analyzing arson samples employing gas chromatography, samples must be prepared for introduction into the instrument. Various sample preparation techniques have been developed differing in complexity and specificity. Some techniques are more efficient for high molecular weight compounds (adsorption-elution techniques and solvent extraction), while others favor compounds of lower molecular weight (headspace techniques). In this experiment, you will be conducting both head-space and solvent extraction preparation techniques for accelerant identification using gas chromatography.

Another means of presumptively identifying potential arson accelerants is to exam their burn characteristics. Some accelerants burn "hot" and/or slowly, extensively charring their surroundings, while others may burn quickly and will not as easily ignite secondary objects.

Purpose

To introduce the student to two sample preparation techniques used for arson analyses (i.e., head-space and solvent extraction) when employing gas chromatography. The student will also become familiar with comparative burn rates and burn patterns for two accelerants (e.g., gasoline and kerosene).

Materials

Obtain the following materials:
1. Small piece of carpet
2. Two watch glasses
3. Two pieces of paper
4. Two 1-dram vials (one with a septum-containing cap)
5. 1 inch square piece of filter paper

The following materials and chemicals will be supplied as needed:
1. Carbon disulfide
2. GC syringes (gas tight)
3. Gas chromatograph
4. Several 1-mL disposable plastic pipettes
5. Accelerants (e.g., gasoline and kerosene)
6. Matches
7. Forceps

Procedures

Part A Preparation of Arson Samples (You may need to work in groups of three or four.)

1. Place 20–30 strands of carpet (including the backing) into a watch glass. Under the hood, add 1½ mL of unknown accelerant to the carpet sample.

2. In a separate fume hood, light the accelerant-soaked carpet sample and allow it to burn for 60 seconds or until the fire goes out (whichever comes first). If the fire does not self-extinguish after 60 seconds, cover the sample with a watch glass. **Record** the results of the burn. (Did it catch fire quickly? Did it burn for the complete 60 seconds? What was the appearance of the flame and smoke?)

3. **Record** the appearance of the charred debris. Remove half of the debris from the watch glass using forceps and place into a 1-dram vial. Place the other half into a separate 1-dram vial. Tightly screw one vial with a septum-containing cap; to the other, screw a normal cap.

4. Set the vial with the septum-containing cap aside. In the fume hood, add 2 mL of carbon disulfide to the second vial and tightly screw down the cap.*

   * Carbon disulfide is fairly noxious and must be handled under the hood. Avoid smelling the chemical and any direct contact.

5. Shake the vial containing the carbon disulfide vigorously for 15–20 seconds. **Record** the appearance of the sample and set it aside until it is your turn to run the sample on the GC/MS.

Part B GC/MS Analysis of Arson Samples

1. Obtain the vial with the septum-containing cap. Remove the cap and insert the needle of a gas-tight syringe into the headspace immediately above the debris. Carefully remove 50–100 μL of headspace.*

   * Gas-tight syringes are very fragile and can be broken easily. EXTREME care must be taken when introducing and dispensing sample; the plunger or needle may bend and destroy the syringe!

2. Allow the instructor to prepare the GC/MS. Very **carefully** dispense all of the collected headspace found in the syringe into the injector of the GC/MS.

3. Wait for, and print out, your results (~15 min).

4. Obtain your vial containing debris and carbon disulfide.

5. Uncap the vial in the fume hood and pour a small amount of carbon disulfide solution from the vial onto a small piece of filter paper. Hold the filter paper such that a corner is directed downward. Allow the carbon disulfide solution to flow to this corner. Gently tap a clean watch glass with the saturated corner of the filter paper.*

   * The above procedure is performed to purify the carbon disulfide solution. We do not want to inject a dirty sample with solid components of debris into the gas chromatograph. Make sure that the drop of carbon disulfide solution tapped onto the watch glass is clear and free of debris.

6. Using a GC syringe (10-μL syringe, does not need to be gas tight), acquire 1–2 μL of carbon disulfide solution from the drop on the watch glass.*

   * Carbon disulfide is very volatile and will evaporate very quickly if it is not introduced into the syringe. Have the instructor quickly examine the drop before introducing it into the syringe. The same above precautions must be followed using the smaller volume GC syringe.

7. Allow the instructor to prepare the GC/MS and then inject the carbon disulfide solution into the GC injector.

8. Wait for your results (~15 min). Print out your results. Acquire the results for accelerant standards prepared by the instructor prior to lab. You will use the results from the standards and your unknowns to answer worksheet questions. Turn in results with your worksheet.

Part C Accelerant Burn Rates and Patterns

1. Obtain one sheet of paper. Crumple it into a **tight** ball. Using a dropper, in the hood, dispense 1–2 mL of gasoline onto the surface of the paper ball. Light the ball in the fume hood shortly after applying the accelerant in order to minimize evaporation. **Record** results of the burning (i.e., charring patterns, rate of ignition, length of burn, whether or not the paper caught fire or just the accelerant burned, rate of solvent evaporation/consumption, etc.).

2. Obtain a second piece of paper. Lay it flat in the fume hood and fold up one of the corners. Using a dropper, in the hood place 1–2 mL of gasoline in an X across the entirety of the paper. Make sure the paper remains flat. Light the X with a match at the bent corner. **Record** results of the burning (i.e., charring patterns, rate of ignition, length of burn, whether or not the paper caught fire, rate of solvent evaporation/consumption, etc.).

3. Repeat steps 1 and 2 using kerosene.

4. Clean up when finished.

Name:_____   Date:_____

Experiment 8 Worksheet

Results and Observations

Part A Preparation of Arson Samples

- o   Observations of burning carpet

- o   Did the carpet self-extinguish?

- o   Appearance of charred debris

- o   Appearance of carbon disulfide charred sample

Part C Accelerant Burn Rates and Patterns

- o   Observations of burning gasoline (ball and flat paper)

- o   Observations of burning kerosene (ball and flat paper)

Experiment 8 Questions

1. What is the number of your unknown and its identification based on the data acquired?

2. Support your answer to #1 by discussing your rationale. (Reference both standard and unknown chromatograms, identifying particular peaks and their matching positions; intensity of peaks is not important for this comparison.)

3. Why were, or could there be, extraneous peaks present in the unknown sample that were/are not found in the known sample?

4. Why is it important that the conditions of the gas chromatograph not be altered to improve the separation of certain components when comparing one accelerant to another?

5. Why did the paper lying flat not burn as well as the crumpled paper regardless of what accelerant was used?

Experiment 9
Explosives

Introduction

All explosive materials contain chemicals that when reacted generate a large amount of hot-expanding gases in a very (very) short amount of time. These hot expanding gases cause the devastating effects associated with explosions. Explosives are classified as either "high" or "low." High explosives have a velocity of detonation greater than 1000 m/s and typically require a blasting cap or a primer (a small amount of sensitive explosive material resulting in a shock). A shock is required for detonation of a high explosive. Many high explosives will not detonate under a flame, but rather will burn slowly. Low explosives have a velocity of detonation less than 1000 m/s and are said to deflagrate (burn rapidly) rather than detonate. In order for a low explosive to explode, it must be confined. Confinement allows gas pressure within the container to increase; an explosion occurs only when the container ruptures.

Explosives may be solids, liquids, or gases. Some explosives may also be gels or emulsions. The majority of explosive materials found at a crime scene are incorporated into improvised explosive devices (IEDs). IEDs are *not* made commercially or militarily, but rather are designed and constructed by individuals in a wide variety of manners. A typical design may include a low explosive such as black powder or smokeless powder confined in a sturdy pipe, tube, or other container and fitted with an ignition system (i.e., a spark, a fuse, or a heating element); although, many other designs employing low and high explosives have been constructed.

Many explosives are analyzed by performing tests that identify the presence of one or more of their components. Black powder is made of carbon, sulfur, and potassium nitrate. Potassium nitrate is very soluble in water and can be removed from black powder through a water extraction; this dramatically decreases the burn rate of the carbon and sulfur left behind. The major component of all smokeless powder samples is nitrocellulose. Nitrocellulose can be extracted from smokeless powder using acetone and subsequently precipitated in water. The precipitate appears as a stringy (dead skin–like) precipitate.

Purpose

To introduce the student to three common low explosives (i.e., black powder, smokeless powder, and sugar/chlorate). Students will become familiar with wet-chemical analysis techniques employed by forensic scientists for analysis of these three low explosives.

Materials

Obtain the following materials:
1. Two 50-mL beaker
2. One100-mL beaker
3. Three 4-inch square pieces of filter paper
4. Spatula

5. Nichrome wire
6. Five 4-inch test tubes
7. Watch glass
8. Spot plate
9. Test tube rack

The following materials and chemicals will be supplied as needed:
1. Black powder
2. Smokeless powder
3. Powdered sugar
4. Potassium chlorate
5. Diphenylamine
6. Concentrated sulfuric acid
7. Methanol
8. Acetone
9. Distilled water
10. Weigh boats
11. Several 1-mL disposable plastic pipettes

Procedures

Part A Analysis of Black Powder

1. Fill a 4-inch test tube with black powder ¼ inch from the bottom. Add 3 mL of distilled water to the test tube. Shake the test tube vigorously for 1 minute while capping the top. Note the appearance of the black powder before and after the addition of water.

2. Fold a piece of filter paper making a funnel and place it on top of a 50-mL beaker. Wet the filter paper so that it stays in place. Pour the aqueous black powder solution over the filter paper placed on top of the beaker and allow all of the water solution to pass through the paper into the beaker. Make sure all of the black soot is transferred to the filter paper. You may need to add a couple additional milliliters of water to the test tube to transfer all of the black soot.

3. Pour the filtered water solution (i.e., the filtrate) into a new 4-inch test tube and place it upright in a test tube rack. Rinse the beaker with distilled water, place the filter paper back into the beaker and wash the black soot again with 5–20 mL of distilled water. Make sure that the sample is well filtered. Collect the second filtrate in a separate test tube. Place aside the filter paper containing the black soot; it will be used in step 7.

4. Weigh out 0.03 g of diphenylamine and place it into a 4-inch test tube. Add to the test tube 3 mL of concentrated sulfuric acid using a 1-mL dropper.* Hold the test tube with one hand and tap the bottom until all of the diphenylamine is dissolved.

* Sulfuric acid is a very dangerous and corrosive acid. If a drop falls from the dropper clean it up immediately. If your clothing or skin comes in contact with the acid, rinse immediately and thoroughly.

5. Test for nitrates: Place 1–2 drops of the second filtrate into a well of a spot plate. Add a small amount of distilled water to dilute the drops. Place 1–2 drops of the diphenylamine solution into the same well of the spot plate. Record the immediate and subsequent results (i.e., color change). Also test the dry powder directly for nitrates by placing a small amount of black powder (a few flakes) into a spot plate and adding a few drops of diphenylamine solution onto the powder. View the black powder flakes and the ring around the murky solution. **Record** your results.

6. Test for potassium ion: Obtain a Bunsen burner and approximately 8 inches of nichrome wire. Bend a small loop into the nichrome wire at one end. Light your Bunsen burner. Clean and dry nichrome wire with distilled water. Wet your flame test loop with distilled water and place it in the flame of the Bunsen burner for 20–30 seconds. **Record** the change of the flame color as the water evaporates and the wire begins to glow. Next, wet your wire with the first filtrate and perform the same test. **Record** your results.

7. Test for carbon: Retrieve your filter paper. Collect a small amount of black soot from the paper with your spatula or tongs. Place the spatula or tongs into the flame of a Bunsen burner for 20–30 seconds or until the soot begins to glow. Remove the spatula from the flame and blow on the glowing soot. **Record** your results.

8. Filter a smokeless powder sample *one time* as you did black powder in steps 1 and 2. Test the smokeless powder filtrate and a dry smokeless powder sample using your diphenylamine solution. **Record** you observations.

Part B Analysis for Nitrocellulose

1. Fill a 4-inch test tube with smokeless powder ¼ inch from bottom. Add 2 mL of acetone to the test tube. Shake the test tube vigorously for 10 seconds while capping the top. Allow any solid powder to settle to the bottom of the test tube.

2. Test for nitrocellulose: Fill a clean test tube approximately two-thirds full with distilled water. Remove some of the acetone from the test tube prepared in step 1 using a plastic pipette and place 2–5 drops into the water-filled test tube. Shake the test tube gently and hold against a dark background. **Record** your results.

Part C Analysis of Burned Samples

1. Acquire a pinch of smokeless powder and place it into a watch glass. Place the watch glass in the hood and light the powder. **Record** your results.

2. Place a small amount of water onto the residue. Mix the water well with the sample by sucking it back into the pipette several times. Place a few drops of the water extract into a spot place. Add a few drops of diphenylamine solution to the extract. **Record** both immediate and subsequent results.

3. Burn a second smokeless powder sample. Test the residue for nitrocellulose by adding a few drops of acetone to the residue using a dropper. Mix the acetone well with the sample by collecting it back into the pipette several times. Add a few drops to a test tube containing water as you did in step 2 of part C. **Record** your results.

4. Repeat steps 1 and 2 for black powder. **Record** your results.

Part D Analysis of Sugar/Chlorate Explosive

1. Measure the mass of approximately 0.1 g powdered sugar and 0.4 g potassium chlorate. Place both onto a watch glass and mix together thoroughly. This will require careful folding of the powder. **Record** its appearance.

2. After the powders are thoroughly mixed, remove approximately 0.01 g and place into a test tube. Add approximately 5 mL of distilled water to the test tube.

3. Test for chlorates: Add one drop of the above solution to a spot plate well. Add one drop of diphenylamine solution to the same well and record both initial and subsequent results.*

   * You may notice that the diphenylamine test can be used to test for both nitrates and chlorates.

4. Make a pile at the bottom of the watch glass with the remaining sugar/chlorate powder.

5. In a fume hood, place 2–3 drops of concentrated sulfuric acid approximately 1 cm from the powder. Shut the sash of the fume hood and allow the sulfuric acid to slowly flow towards the pile of sugar/chlorate powder.* **Record** your observations.

   * It is important to shut the sash since the initiation of the reaction may cause spattering. First you will see some bubbles and then you should see a flame. It may take several seconds for ignition. Plan where you will be placing the sulfuric acid dropper ahead of time (leave in the fume hood).

6. Test for chlorates: After ignition, add 3–5 mL of distilled water to the residue in the watch glass. Mix the water well with the sample by sucking it back into the pipette several times. Add one drop of this distilled water solution to a spot plate

well. Add one drop of diphenylamine solution to the same well and **record** both initial and subsequent results.

7. Clean up when finished.

Name:_____   Date:_____

Experiment 9 Worksheet

Results and Observations

Part A Analysis of Black Powder

- o Appearance of black powder before and after addition of water

- o Observations of diphenylamine test on filtrate

- o Observations of diphenylamine test on dry sample

- o Observations of flame test with water

- o Observations of flame test with black powder filtrate

- o Observations of carbon test

- o Appearance of smokeless powder before and after addition of water

- o Observations of diphenylamine test on smokeless powder filtrate

- o Observations of diphenylamine test on dry smokeless powder sample

Part B Analysis for Nitrocellulose

o   Observations of nitrocellulose test using smokeless powder

Part C Analysis of Burned Samples

o   Observations of burning smokeless powder

o   Observations of diphenylamine test using burned smokeless powder

o   Observations of nitrocellulose test using burned smokeless powder

o   Observations of burning black powder

o   Observations of diphenylamine test using burned black powder

Part D Analysis of Sugar/Chlorate Explosive

o   Observations of sugar/chlorate powder mixture

o Observations diphenylamine test using unburned sugar/chlorate powder in distilled water

o Observations of the burning of sugar/chlorate powder

o Observations of diphenylamine test using burned sugar/chlorate powder in distilled water

Experiment 9 Questions

1. Why didn't the black soot in part A for black powder ignite and burn as quickly as black powder should? Could moisture cause problems when using black powder? Explain.

2. Discuss the differences seen between the diphenylamine test results for the burned and unburned smokeless and black powder filtrate samples. Theorize as to why the results differ for each sample.

3. Discuss how you could distinguish smokeless, black, and sugar/chlorate powders based on appearance, burn characteristics, and chemical testing.

Experiment 10
Identification of Blood

Introduction

Blood, the cherished fluid of life, is unique in appearance, physical characteristics, and composition. Wet bloodstains typically have a characteristic brilliant red color, while dried bloodstains often appear dark red to brown in color. Blood has unique cohesive properties that allow it to adhere to, and stain, almost anything. The composition of blood, although dynamic and influenced by health, is unique, containing red blood cells, white blood cells, platelets, proteins, and minerals exclusive to blood.

Although the appearance of blood is distinctive, many red stains made by other substances (e.g., ketchup, paint, jam, etc.) may be strikingly similar. Because of this, tests have been developed that allow forensic scientists to quickly and presumptively identify a stain as blood through chemical exploitation of blood's unique composition. Most tests use the heme portion of hemoglobin, found only in the red blood cells of blood, to catalyze a reaction that results in either a color change or the generation of light. All of these tests are quite sensitive (i.e., a very faint blood stain invisible to the naked eye can be located and identified), yet only presumptive in nature. Unfortunately, many other substances may also generate the indicative results (e.g., rust, potatoes, bread) and generate a false positive. However, these substances do not typically share the same unique appearance of blood; yet, a mixture of one of these compounds with something having the appearance of blood could become misleading. It is important to be aware of these substances such that a complete and accurate interpretation of results may be obtained.

Purpose

To introduce the student to common blood identification tests, their sensitivity, and results. Students will become familiar with various substances that potentially give false positives for the presence of blood. Students will recognize the presumptive nature, and limitations, of these tests.

Materials

Obtain the following materials:
1. Five 1-mL disposable plastic pipettes
2. One 100-mL beaker
3. Three 50-mL beakers
4. Filter paper

The following materials and chemicals will be supplied as needed:
1. Spatula
2. Weigh boats
3. Gloves
4. Dark room

5. Leucomalachite green
6. Sodium perborate
7. Glacial acetic acid
8. Distilled water
9. O-tolidine
10. Hydrogen peroxide
11. Ethanol
12. 3-aminophthalhydrazide (luminol)
13. Sodium carbonate
14. Sodium perborate
15. Whole dog blood (or sheep)
16. Bleached water
17. Soapy water
18. False positive samples (e.g., apple, cabbage, blackberry jam, bread, horseradish, tomato, rust, potassium permanganate)

Procedures

Part A Preparation of Solutions and Samples

1. Prepare 5.0 mL of the following presumptive blood identification solutions in a 50-mL beaker.

    A) Leucomalachite Green Solution
Sodium perborate	0.16 g
Leucomalachite green	0.005 g
Glacial acetic acid	3.3 mL
Distilled water	1.66 mL

    B) O-tolidine Solution
       Solution 1 (You will add this first when testing.)
O-tolidine	0.08 g
Ethanol	2.0 mL
Glacial acetic acid	1.5 mL
Distilled water	1.5 mL
       Solution 2 (You will add this second when testing.)
       3% hydrogen peroxide

    C) Luminol Solution (This solution will take 20–30 min to dissolve completely. You may need to use the mechanical stirrer.)
3-aminophthalhydrazide	0.005 g
Sodium carbonate	0.025 g
Sodium perborate	0.035 g
Distilled water	5.0 mL

2. Prepare the following solutions of 10%, 1%, 0.1%, 0.01%, 0.001%, 0.0001% animal blood in distilled water. To prepare the first solution, mix one drop of animal blood with nine drops of distilled water. Carryout serial dilutions to prepare the next five solutions.

3. Prepare the following samples in triplicate by rubbing a generous amount of each sample onto a piece of filter paper: apple, cabbage, blackberry jam, bread (may need to place on paper), horseradish, tomato, rust, and one drop potassium permanganate solution.

Part B Analysis of Samples and Blood Dilutions

1. Test the sensitivity of your solutions prepared in part A. Add one drop of each blood dilution prepared in part A in triplicate to a piece of filter paper. Next, test each presumptive blood identification solution. **Record** your results.

2. Test the samples (other than blood) prepared above using all three of the solutions prepared in part A. Add one drop of solution to the filter paper containing the sample and **record** both the immediate and subsequent results. A reaction will not occur between solution B and a sample until one drop of hydrogen peroxide is subsequently added. Tests with luminol must be done in a dark room.

Part C Analysis of Cleaned Bloodstains

1. Add one drop of the most dilute blood solution that gave a positive test for luminol to a piece of filter paper. Add a second drop to a separate piece of filter paper. To one piece of filter paper, add one drop of soapy water and to another add one drop of bleach. Allow the pieces of filter paper to dry.

2. Test the filter paper above using luminol. **Record** your results.

3. Clean up when finished.

Name:_____   Date:_____

Experiment 10 Worksheet

Results and Observations

Part B Analysis of Samples and Blood Dilutions

Samples

	Leucomalachite Green	O-tolidine	Luminol
Apple			
Cabbage			
Blackberry Jam			
Bread			
Horseradish			
Tomato			
Potassium Permanganate			
Rust			

Blood Dilutions

	Leucomalachite Green	O-tolidine	Luminol
10%			
1.0%			
0.10%			
0.010%			
0.0010%			
0.00010%			

Part C Analysis of Cleaned Bloodstains

o Observations of bloodstain with soapy water added

o Observations of bloodstain with bleached water added

Experiment 10 Questions

1. Which samples if mixed with something having the appearance of blood could give ambiguous results for the presumptive identification of blood? Explain.

2. The o-tolidine test used in this experiment requires two reagents. Both reagents must be added for a positive identification. Sometimes a color change will occur after adding solution 1, but before adding solution 2 (e.g. as with potassium permanganate); this is a false positive. Discuss how this two-step test has an advantage over the other tests.

3. Which test solution was the most sensitive (test positive with least amount of sample)? Explain.

4. Luminol reacts with bleach. Did the addition of the cleaning solutions alter the sensitivity of luminol to the blood samples? Explain.

5. Because hemoglobin is only a catalyst to the reactions, is it possible, given enough time, that everything tested would result in the development of a blue-green color indicating the presence of blood? Explain.

Experiment 11
Fingerprints

Introduction

Ever since the early 1900s, fingerprints have been successfully used to solve crimes. The success has been accredited to fingerprints' ability to identify a suspect at the exclusion of all others. No two fingerprints have ever been found alike. In addition, during the course of one's activities, fingerprints are left on several types of surfaces. Many processing techniques have been developed allowing fingerprints to be enhanced on many of these surfaces. Each technique has the ability to selectively *interact,* or chemically *react,* with a component, or several components, found within the fingerprint residue. However, since fingerprint residue can be composed of many different substances (amino acids, water, salts, dust, oil, etc.), and the composition of which is related to each individual's body chemistry and activities, not all techniques work the same on all prints. Also, not all surfaces can be processed for fingerprints. Many rough surfaces tend to be extremely challenging. Furthermore, the quality of the print may be poor or damaged due to the manner in which the print was left and/or due to subsequent activities (e.g., brushing, washing, rain, etc.). Despite these potential disadvantages, fingerprints continue to be highly recognized as extremely valuable evidence.

Purpose

To introduce the student to various fingerprint development techniques used on both porous (i.e., paper) and nonporous (i.e., glass) surfaces. The student will recognize many advantages and disadvantages of each technique. It should also become apparent to the student that not all techniques work the same on all prints.

Materials

Obtain the following materials:
1. 250-mL beaker
2. One watch glass
3. One weigh boat

The following materials and chemicals will be supplied as needed:
1. Ultraviolet light source
2. Flash light
3. Fingerprint brushes
4. Black and fluorescent fingerprint powder
5. Lifting tape
6. Iodine crystals, 600-mL beaker, and a hot plate
7. Ninhydrin solution, small tray, tongs, and an iron
8. Superglue
9. Oven
10. Silver nitrate solution and a small tray

11. Crystal violet solution and a small tray

Procedures

Part A Fingerprints and Light Sources

1. Prepare two fingerprints, approximately two inches apart, of the same finger on both a piece of paper and a watch glass (a total of four prints). Good prints can be prepared by first rubbing your fingers on your forehead or some other oily or sweaty part of the body. Next, gently press the finger onto the surface. Do not press too hard or the ridges will smash together and a poor fingerprint will result.

2. Using oblique lighting, attempt to find the prints made on each surface. Oblique lighting can be obtained by shining a light source at an angle to the surface or rotating the surface in a fixed light source. **Record** your results.

3. Using an ultraviolet light source in the darkroom, attempt to find your prints on each surface. Some fingerprint residue may have inherent fluorescence. Others may fluoresce due to residue acquired (e.g., cosmetic products). **Record** your results.

Part B Fingerprint Powder

1. Obtain a little black and fluorescent fingerprint powder and two fingerprint brushes (one for each powder used). Develop one print on each surface from part A using black powder, and one on each surface using fluorescent powder.*

    * Only use fluorescent powder brushes with fluorescent powder. You may wish to do this over a sink and allow extra powder to slowing fall down the paper or watch glass adhering to your prints. Do not brush too hard or you may damage the prints. Many brushes may already have enough powder sufficient for enhancement and will not require dipping into the powder.

2. View both fingerprints under normal and ultraviolet light. **Record** your results. If pattern type cannot be determined from the prints (i.e., arch, loop, or whorl), repeat part A just one more time. Review your text and/or the Internet to become familiar with these pattern types.

3. Lift the black print from the watch glass using Scotch® tape (or any other clear tape) by carefully placing a piece of tape over the print, gently rubbing, and then removing. Place the "lift" on your worksheet. Attempt to re-dust the print and lift again. Place the second "lift" on your worksheet. Wash your watch glass with soap and water removing all fingerprints.

Part C Iodine Fuming

1. Prepare two prints as in part A on a piece of paper using a *different* finger.

2. Make sure the piece of paper is large enough to fit over the mouth of the beaker on the hot plate found in the fume hood. While wearing gloves, obtain 2–3 iodine crystals using a spatula and place them into the hot beaker. Immediately place the paper containing the prints over the mouth of the beaker with the prints down and allow the iodine fumes to fumigate the paper. Watch the prints develop through the side of the beaker. **Record** your results.

3. Place a piece of tape over the top and bottom of the paper where one of the prints is found and add a few drops of starch solution over the second.* Allow the paper to dry and then tape it on your worksheet. If the print type is not legible, repeat part C just one more time.

   * Iodine-fumed prints fade over time. Placing a piece of tape over the top and bottom of the print will decrease fading. Alternatively, the prints may be fixed with starch solution.

Part D Ninhydrin Development

1. Prepare two prints as in part A on a piece of paper using again a *different* finger. Note: A sweaty print may be better than an oily print for this part.

2. While wearing gloves, dip the piece of paper into the ninhydrin solution very briefly. The ninhydrin solution is flammable. Do not expose to flame or high heat. The solution will also temporarily change your skin color in the locations in which it comes in contact. Caution must be taken.

3. Using tongs, fan dry the paper. Steam the paper with an iron for a few minutes without having the iron touch the surface of the paper.

4. If prints have not developed, turn off the steam and hold the iron on the surface of the paper for a few seconds until the prints begin to develop. **Record** your observations.

5. Cut out your prints and tape them on your worksheet. If the print type is not legible, repeat part D just one more time.

Part E Superglue Development

1. Obtain a 250-mL beaker, a watch glass, a weigh boat, and some superglue. Place 5–10 mL of water in the bottom of the beaker. Each partner make one fingerprint, of yet **another** finger, somewhat side-by-side and close to the center of the watch glass. Each partnership will use one beaker. You may need to mark whose print is

whose. Using oblique lighting, make sure that you can see each fingerprint and that they are of a good quality.

2. Place the weigh boat in the beaker containing water. Add a few drops of superglue into the weigh boat and immediately place your watch glass containing the prints over the top of the beaker. Place the beaker in the oven for 10–20 minutes.*

   * Check on your prints periodically. It is possible to over-fume your prints and have too much superglue adhere to the prints and watch glass, destroying detail.

3. Remove the beaker. **Be careful— it is hot.** Allow the beaker to cool. Remove the watch glass and examine your prints under oblique lighting. **Record** the appearance of the superglued prints.

4. Dust the superglued fingerprints using black fingerprint powder and "lift" them using tape. Attempt to re-dust the prints and "lift" again. Place both "lifts" on your worksheet. Wash your watch glass with soap and water removing all fingerprints. You may need to use acetone to remove the superglue.

Part F Silver Nitrate Development

1. Prepare two prints as in Part A on one piece of paper using a *different* finger. Note: A sweaty print is better than an oily print for this part.

2. Dip the paper into a tray containing silver nitrate solution using tongs. Remove the paper and let drip any excess from the paper into a waste container.

3. Set the paper near the window or an ultraviolet light source and allow it to develop (15–20 minutes). Cut and tape your prints on the worksheet. If the prints are not legible, repeat part F one more time.

Part G Sticky-side of Tape Development

1. Obtain a piece of Scotch® tape large enough for two fingerprints. Make two prints on the sticky side of the tape.

2. Soak the piece of tape in the crystal violet solution for a few minutes. Remove using tongs and rinse using distilled water.

3. Examine the contrast. If the prints are faint, soak the tape for a little longer and then rinse again.

4. Allow the tape to dry and tape or staple the developed prints onto your worksheet. If prints are not legible, repeat part G just one more time.

Part H Unknown Prints

1. Using the development technique of your choice, develop the unknown prints found a piece of paper supplied by the instructor.

2. Place the developed unknown prints on your worksheet.

Part I Rolling Prints (Optional)

1. Allow the instructor to prepare the inking station.

2. Practice rolling your partners fingerprints onto a 10-print card. Roll all 10 fingers. Roll prints into the center of the body. Remember to hold both ends of the finger. Turn in 10-print card of your partner with your worksheet.

3. Clean up when finished.

Name:_____   Date:_____

Experiment 11 Worksheet

Results and Observations

Part A Fingerprints and Light Sources

o   Observations of fingerprints on both surfaces using oblique lighting

o   Observations of fingerprints on both surfaces using ultraviolet lighting

Part B Fingerprint Powder

o   Observations of fingerprints on each surface using black powder
    - Normal lighting:

    - Ultraviolet lighting:

o   Observations of fingerprints on each surface using fluorescent powder
    - Normal lighting:

    - Ultraviolet lighting:

o   Place first and second black powder "lifts" from the watch glass below and identify the general pattern type (i.e., arch, loop, or whorl). Note: For all prints turned in, if the prints are not legible, view your finger and identify what it should be.

Part C Iodine Fuming

o Observations of fingerprints upon being fumed

o Observations of print with starch solution

o Place both taped and starched iodine prints below and identify the general pattern type.

Part D Ninhydrin Development

o Observations of fingerprints developed using ninhydrin

o Place ninhydrin-developed prints below and identify the general pattern type.

Part E Superglue Development

o Observations of superglued fingerprints

o Place both "lifts" below and identify the general pattern type.

Part F Silver Nitrate Development

o Observations of silver nitrate development

o Place silver nitrate developed prints below and identify the general pattern type.

Part G Sticky-side of Tape Development

o Observations of sticky-side of tape development

o Staple or tape developed tape below and identify the general pattern type.

Part H Unknown Prints

o Observations of unknown print developments

o Place developed unknown prints below and identify the general pattern type.

Experiment 11 Questions

1. Explain why or why not part A may serve as a good initial technique.

2. Why do the iodine crystals need to be heated in part C?

3. According to your results in part E, did superglue fuming do much enhancing of the fingerprint? Why did we superglue fume?

4. Did each technique work the same for you and your partner? Explain.

5. Discuss why you chose the development technique employed to develop the unknowns. Make comparisons to other techniques, discussing advantages and disadvantages. If your first choice did not work, explain your results and what you did next.

Experiment 12
Heat Capacity and Fire Fighting

Introduction

Specific heat capacity describes the amount of energy (heat) required to raise the temperature of one gram of substance one degree Celsius. It can also be used to describe the amount of heat leaving (or being transferred from) one gram of substance as it cools one degree Celsius. Heat may be expressed in units of calories (cal) or Joules (J). For example, the specific heat capacity of water is 1.00 cal/g °C or 4.18 J/g °C. Most substances have a specific heat capacity less than that of water.

A calorimeter is a device used to measure the amount of heat produced or transferred during a chemical or physical process. These devices can be used to determine the specific heat capacity of a substance. Calculations are based upon the conservation of energy (the first law of thermodynamics), i.e., energy is neither created nor destroyed during a chemical or physical process. In this experiment, a calorimeter will be used to measure the specific heat capacity of steel and concrete.

Prior to measuring the specific heat capacity of a substance with a calorimeter, the *calorimeter constant* must be determined. This constant describes the amount of heat absorbed or lost by the entire mass of the calorimeter during the measurement of heat transfer expressed as a temperature change. For example, assume we add hot water to a calorimeter containing cold water. The heat from hot water will be transferred to the cold water and also to the calorimeter, raising the temperature of both. The heat lost by the hot water will reduce its temperature until all three objects (i.e., the calorimeter, the cold water, and the hot water) are in thermal equilibrium. This can be represented mathematically with the following expression.

$$Q_{HotWater} = -\left(Q_{ColdWater} + Q_{Calorimeter}\right)$$ 12-1

Q is used to represent heat. The negative sign is used to indicate that the heat lost by the hot water is gained by the cold water and the calorimeter. Each Q can also be expressed as

$$Q = sm\Delta T$$ 12-2

where s is the specific heat capacity of the substance, m is its mass, and ΔT is its change in temperature (i.e., final temperature minus initial temperature). Typically, for the calorimeter heat expression, the calorimeter constant, C_{cal} is the product of the calorimeter's mass and its specific heat capacity (i.e., $C_{cal} = sm$). Thus, we can rewrite equation 12-1 as the following:

$$sm\Delta T_{HotWater} = -\left(sm\Delta T_{ColdWater} + C_{cal}\Delta T\right)$$ 12-3

In this experiment, the above equation will be used to determine the calorimeter constant of a Styrofoam cup calorimeter with all other variables being known. A known mass of hot water will be mixed with a known mass of cold water in the calorimeter and the change in temperatures will be recorded. Water has a density of approximately 1.00 g/mL at room temperature, thus a volume of water will be measured using a graduated cylinder rather than mass. The temperature change of the cold water and the calorimeter will be the same (i.e., cold to warm), while the temperature change of the hot water will be hot to warm.

If the hot water is replaced with a hot object of unknown specific heat (e.g., steel or concrete), equation 12-3 can be rewritten as the following:

$$sm\Delta T_{HotObject} = -(sm\Delta T_{ColdWater} + C_{cal}\Delta T) \qquad 12\text{-}4$$

After determining the calorimeter constant, equation 12-4 may be used to calculate s of an object of unknown specific heat capacity.

Fire fighters typically use water as their greatest resource when fighting fires. Water quenches fire by reducing its temperature. For example, wood has an ignition temperature of approximately 200 °C; below this temperature wood will not burn. The addition of enough water to reduce its temperature to 200 °C or below will abate the fire. However, if the fire is burning in a cement-framed building, it is not just the wood (or fuel) that needs to be cooled, but the entire cement building itself. Water cools the fire by transferring the heat of the cement and fuel to itself as it rises in temperature. Because water has a relatively high specific heat capacity, it can reduce the temperature of most substances more dramatically than its temperature will rise. Water additionally carries away a significant amount of heat as it is converted from a liquid to steam. However, if the water is not turned into steam (which is **not** typically the case), the amount of water needed to reduce the temperature of a burning building composed mostly of cement can be determined if the fire fighter has a rough idea of the mass of the building, the average temperature of the fire, and the specific heat capacity of cement.

Purpose

To introduce the student to calorimeters and their measurements, calorimeter constants, and specific heat capacities. Each student will become familiar with the potential application of heat capacity in fire fighting.

Materials

Obtain the following materials:
1. 2 test tubes
2. 600-mL beaker
3. 100-mL beaker
4. Hot plate or ring stand with ring and wire mesh to boil water
5. Bunsen burner
6. Matches or striker
7. Two Styrofoam cups
8. Lids for cups with a hole for a thermometer
9. Two thermometers

10. Two ring stands and clamps (one for each thermometer)
11. 100-mL graduated cylinder
12. Styrofoam cup lid or cardboard square with hole for thermometer
13. Test tube holder
14. Beaker tongs
15. Watch glass

The following materials and chemicals will be supplied as needed:
1. Pebble rocks or cement fragments
2. Steel fragments or ball bearings

Procedures

Part A Determination of the Calorimeter Constant

1. Obtain two dry Styrofoam cups and nest one inside the other. This will increase the ability of your "make-shift" calorimeter to insulate.

2. Measure 100 mL of water using a graduated cylinder and pour it into the Styrofoam cup assembly. **Record** the exact volume of water transferred. Cover the cups with a lid and insert a thermometer to measure its temperature. Periodically check its temperature until it is no longer changing.

3. Begin heating approximately 250 mL of water with a ring stand and Bunsen burner or hot plate. Introduce a thermometer (different from the thermometer used in step 2) and heat the water to approximately 70–80 °C.

4. Remove the water from heat using gloves or tongs and carefully measure 100 mL of hot water in a graduated cylinder. Pour this into a 250 mL beaker, cover with a lid or watch glass, and allow it cool to approximately 60 °C. **Record** the exact volume dispensed.

5. Check the cold water one last time from step 2 and make sure that its temperature has reached equilibrium and is not changing. **Record** its temperature.

6. When the hot water has cooled to a temperature of approximately 60 °C, **record** its exact temperature and quickly (but carefully) transfer the entire volume into the calorimeter containing the cold water.

7. Cover the calorimeter again with a lid and carefully stir the solution with the thermometer through the lid hole until the temperature stops rising. **Record** the highest temperature reached before the temperature begins to fall.

8. Repeat steps 2–7 two more times.

Part B Determination of the Specific Heat Capacity of Concrete and Steel

1. Obtain two test tubes and approximately 25 grams of rock or cement and 50 grams of steel. Make sure the rock and steel samples are small enough to be placed into your test tubes.

2. Make sure the samples are dry, measure and record their masses to ±0.01 g using a balance, and place the steel sample into one test tube and the rock or cement sample into another. The samples should not be filled to the top of the test tubes. There should be 1–2 inches of headspace above the samples.

3. Begin boiling approximately 500 mL of water in a 600-mL beaker. The water level should be high enough to completely submerge the sample in the test tubes, but below the height of the test tube to eliminate the introduction of water into the test tubes during boiling.

4. Allow the samples to remain in the boiling water for approximately 15 minutes or longer.

5. While the samples are in the boiling water, dry your calorimeter. Measure approximately 50 mL of cold (room temperature) water using a graduated cylinder and pour it into the calorimeter. Cover and allow the water to reach thermal equilibrium. **Record** the volume of water dispensed and the temperature of the cold water after it has reached thermal equilibrium.

6. After approximately 15 minutes of boiling, **record** the temperature of the boiling water and remove one of the samples with a test tube holder. Remove the lid of the calorimeter and quickly (but carefully, to avoid splashing) pour the sample into the cold water. Carefully stir the sample and **record** the highest temperature reached before the temperature begins to fall.

7. Repeats steps 5–6 for the second sample.

8. Repeat steps 2–7 two or three more times (ask your instructor).

Name:_____   Date:_____

Experiment 12 Worksheet

Results and Observations

Part A Determination of the Calorimeter Constant

	Trial #1	Trial #2	Trial #3
Volume of Cold Water, mL			
Volume of Hot Water, mL			
Initial Temperature of Cold Water and Calorimeter, °C			
Initial Temperature of Hot Water, °C			
Final Temperature of Everything, °C			

Part B Determination of the Specific Heat Capacity of Concrete and Steel

Steel Sample

	Trial #1	Trial #2	Trial #3
Volume of Cold Water, mL			
Mass of of Hot Steel, mL			
Initial Temperature of Cold Water and Calorimeter, °C			
Initial Temperature of Hot Steel, °C			
Final Temperature of Everything, °C			

Rock or Concrete Sample

	Trial #1	Trial #2	Trial #3
Volume of Cold Water, mL			
Mass of Hot Concrete, mL			
Initial Temperature of Cold Water and Calorimeter, °C			
Initial Temperature of Hot Concrete, °C			
Final Temperature of Everything, °C			

Experiment 12 Questions

1. Using equation 12-3 and your data from part A, calculate the calorimeter constant for each trial in units of J/°C. Remember that the specific heat capacity of water is 4.18 J/g °C. Determine the average calorimeter constant. Show your work.

2. Using the average calorimeter constant value calculated in question 1, equation 12-4, and your data from part B; calculate the specific heat capacity of steel for each trial in units of J/g °C. Determine the average specific heat capacity. Show your work.

3. Using the average calorimeter constant value calculated in question 1, equation 12-4, and your data from part B; calculate the specific heat capacity of rock (cement) for each trial. Determine the average specific heat capacity. Show your work.

4. Based on your results for questions 2 and 3, determine the number of gallons of water (approximately 3790 g equals 1 gallon of water) to abate a fire raging in a 50,000 kg building built mostly of concrete at an average temperature of 800 °C. Assume the water does not turn to steam. The temperature must be reduced to 200 °C. Assume that the water temperature starts at 25 °C and is raised to 100 °C. Repeat the calculation for a steel building under the same conditions. Show your work.

5. In part B, you are told to pour your hot sample carefully into the calorimeter to avoid splashing. What would happen to the calculated specific heat capacity if an unknown amount of water were to splash out of the calorimeter? Explain.

Experiment 13
Bloodstain Pattern Analysis

Introduction

Bloodstains left at a crime scene, for many, present a gruesome and undesirable display of inexplicably horrid events, while to an investigator; these stains open what may be the only window to the past. Bloodstain pattern analysis is the examination of the distribution and shape of bloodstains in an attempt to acquire information associated with the events of their origination. There are many types of bloodstains (e.g., spatter, swipes, wipes, drips, spurts, etc.). It is the task of the bloodstain pattern analyst to interpret the bloodstains and offer some insight into what may have occurred. The origination of a bloodstain, the force involved in an impact, the sequence of events, and information specific to the location of a victim can all be interpreted from bloodstain patterns.

The distance at which a blood droplet fell can be determined by measuring the diameter of its bloodstain. On smooth surfaces, droplet size will increase logarithmically with distance and begin to level out at approximately 4–6 ft as the droplet reaches terminal velocity. Blood droplets must fall at a 90-degree angle to the surface for distance-fallen calculations to be determined. Otherwise, the bloodstain generated will not be a perfect circle having a uniform diameter; the bloodstain will be oval and have both a major and minor axis.

The angle and direction of impact of a blood droplet can be determined by examining the shape and major and minor axes of its bloodstain. The shape of a bloodstain produced at an angle less than 90 degrees to its surface is oval with a tail or tapered end in the direction of impact. For example, the simulated bloodstains in Figure 13.1 originated from a location to the left and struck the surface while traveling from left to right. If the direction of impact is known for several bloodstains, you can trace each back to their common origin (i.e., the point where the lines showing direction cross). You will be performing this determination in part D of this experiment.

Figure 13.1. Simulated bloodstains with droplets originating from the left, traveling from left to right, and striking a surface at an angle less than 90 degrees to the surface. The small bloodstains found close to the first large bloodstain are often called satellite or secondary bloodstains. They are produced after the formation of the larger bloodstain.

The angle of impact can be determined by measuring both the major and minor axes of the bloodstain. Typically, the bloodstain is not a perfect oval due to tails, tapered ends, and satellite droplets. However, in order to determine the length of the major and minor axes, an oval must be assumed. Figure 13.2 shows the first bloodstain from Figure 13.1 with an assumed oval and its major (D) and minor (d) axes. The angle of impact with respect to the surface can be determined by taking the inverse sine of the ratio of the minor axis over the major axis (i.e., $\sin^{-1}$ (d/D)). However, if the surface is at an angle and the blood droplet falls straight down striking the surface, the angle of the surface can be determined by taking the inverse cosine of the ratio of the minor axis over the major axis (i.e., $\cos^{-1}$(d/D)). You will be determining the latter is this experiment.

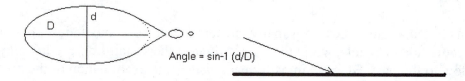

Figure 13.2. Assuming an oval and finding the major and minor axes for angle of impact determination.

Bloodstain spatter is usually generated by a force imparting velocity to the droplets. Low-velocity spatter (less than 5 ft/s) produces bloodstains with most having a diameter greater than 4 mm. Medium-velocity spatter (5–100 ft/s), generated by a violent strike, produces bloodstains with a 1–4 mm diameter. High-velocity spatter (greater than 100 ft/s), generated from a cough or gunshot, produces bloodstains with most having a diameter less than 1 mm. While the directionality and angle of impact of bloodstains can be determined for low- and medium-velocity spatter, the bloodstains of high-velocity spatter are usually too small for such determinations.

Purpose

To generate and analyze various types of bloodstain patterns. The student will generate and analyze drips of simulated blood at various angles and on various surfaces. The student will also generate and analyze medium-velocity spatter. Prior to generating drip patterns, the student will measure the volume of droplets from a dropper and confirm volume consistency.

Materials

Obtain the following materials:
1. Several 1-mL disposable plastic pipettes
2. 100-mL beaker
3. 50-mL beaker
4. Paper
5. Manila folder
6. Glass

The following materials and chemicals will be supplied as needed:
1. Fake blood (e.g., whipping cream and food coloring)
2. Mousetrap
3. Measuring tape
4. Calipers
5. Protractor

6. Six sheets of white paper
7. Tape

Procedures

Part A Determining Droplet Volume

1. Using a 1-mL pipette, count the number of drops found in one milliliter of fake blood. Hold the pipette at a 45-degree angle to the ground and dispense 1 mL of fake blood drop-by-drop into a beaker. The droplet size may change based on the angle of the pipette, so it is important that it is always held at a constant angle. **Record** your results.

2. Repeat step 1 four more times. **Record** your results and calculate the average droplet size in mL.

Part B Stain Size

1. Holding a pipette containing fake blood at a 45-degree angle, dispense three drops of fake blood onto a piece of paper at the following distances: 1, 6, 12, 24, 36, and 48 inches.

2. Using calipers, measure the diameter of each droplet and **record** your results.

3. Repeat steps 1 and 2 for manila folder, glass, and floor.

4. Measure the diameter of droplets generated from an *unknown* distance (the instructor will inform you as to where these droplets are located) taking note of the surface on which they are found. An estimation of the distance from which these droplets fell will be made based on data acquired above. Graphs will be generated for data acquired above and turned in with your worksheet.

Part C Angle of Impact

1. Holding a pipette containing fake blood at a 45-degree angle, dispense a drop of blood onto a manila folder held at a 15-degree angle (use a protractor), with respect to the bench, at approximately 12 and 36 inches. **Record** the width (minor axis, d) and length (major axis, D) of the oval droplet using calipers. Calculate the assumed angle at which the manila folder was held with respect to the bench top using the equation $\cos^{-1}(d/D)$ and the measured major and minor axes. **Record** the calculated angle.

2. Repeat step 1 at the following angles: 45, 55, and 75 degrees.

3. Repeat steps 1–2 for glass.

4. Measure and **record** the major and minor axes of droplets that fell and struck a surface held at an *unknown angle*. The instructor will inform you as to where these droplets are found. **Record** your results.

Part D Medium-Velocity Spatter

1. Obtain a mousetrap and six sheets of white paper. Tape all six sheets together, producing one large sheet. Place the large sheet of paper flat in a hood or in an area where generated spatter is not a concern. If using a fume hood, pull down the sash to reduce back spatter on yourself.

2. Place a small amount of fake blood on the end of the mousetrap where the metal loop will hit. Place the mouse trap at one end of the paper and pull back the loop part way and release it allowing the generation of medium velocity blood spatter.

3. Repeat steps 1 and 2, holding the mousetrap at two different angles. **Record** your observations.

4. Clean up when finished (you will have made a mess).

Name:_____   Date:_____

Experiment 13 Worksheet

Results and Observations

Part A Determining Droplet Volume

	Trial #1	Trial #2	Trial #3	Trial #4	Trial #5
drops/mL					

Average drops/mL:_____

Part B Stain Size

Diameter of each drop

Paper	1 inch	6 inches	12 inches	24 inches	36 inches	48 inches
Trial 1						
Trial 2						
Trial 3						
Average						

Manila F	1 inch	6 inches	12 inches	24 inches	36 inches	48 inches
Trial 1						
Trial 2						
Trial 3						
Average						

Glass	1 inch	6 inches	12 inches	24 inches	36 inches	48 inches
Trial 1						
Trial 2						
Trial 3						
Average						

Floor	1 inch	6 inches	12 inches	24 inches	36 inches	48 inches
Trial 1						
Trial 2						
Trial 3						
Average						

Graph the average of the three trials at each distance for each surface in part B. Attach the graph to this worksheet. Mark on your graphs with a circle where the unknown droplets correspond.

Unknowns

Surface #1:_____ Diameter #1:_____ Distance #1:_____
Surface #2:_____ Diameter #2:_____ Distance #2:_____

Part C Angle of Impact

Manila Folder

15 degrees	12 inches	36 inches	**45 degrees**	12 inches	36 inches
Major axis (D)			Major axis (D)		
Minor axis (*d*)			Minor axis (*d*)		

55 degrees	12 inches	36 inches	**75 degrees**	12 inches	36 inches
Major axis (D)			Major axis (D)		
Minor axis (*d*)			Minor axis (*d*)		

Glass

15 degrees	12 inches	36 inches	**45 degrees**	12 inches	36 inches
Major axis (D)			Major axis (D)		
Minor axis (*d*)			Minor axis (*d*)		

55 degrees	12 inches	36 inches	**75 degrees**	12 inches	36 inches
Major axis (D)			Major axis (D)		
Minor axis (*d*)			Minor axis (*d*)		

Manila F.	Cal. An. 12 in	Cal. An. 36 in	**Glass**	Cal. An. 12 in	Cal. An. 36 in
15 degrees			15 degrees		
45 degrees			45 degrees		
55 degrees			55 degrees		
75 degrees			75 degrees		

Unknowns

	Unknown #1	Unknown #2
Major axis (D)		
Minor axis (*d*)		
Angle of Impact		

Part D Medium-Velocity Spatter

o **Record** your observations.

Experiment 13 Questions

1. On your medium-velocity spatter, circle five droplet stains that correspond to one impact site. Draw lines through the center of the droplets back to the origin. Do this for each of the three origins of blood spatter. Only one person in each group needs to turn this part in.

2. When is it important in bloodstain pattern analysis that blood droplet size remains relatively constant? Why?

3. Based on the analyses performed in this lab, could someone determine both angle of impact and distance fallen from a single bloodstain? Explain.

4. Would results differ if a different liquid was used instead of fake blood? Explain.

Experiment 14
Glass

Introduction

Glass is ubiquitous, being found in virtually every home, business, and automobile. Due to its prevalence (and to it often being broken during the commission of a crime) it is frequently found as evidence at a crime scene either loose or adhering to other items of evidence (e.g., suspect's clothing). Glass varies in composition due to the usage of different raw materials and/or additives. These variations in composition aid in glass comparisons. Glass also may vary due to post-manufacture treatment (e.g., tempered glass).

Glass evidence is typically considered class evidence. Rarely can an analyst conclude that two fragments of glass share a common origin at the exclusion of all other potential origins. Glass evidence is examined for comparative properties by probing its physical and chemical make-up. Glass is also often examined to determine the manner in which it was broken. This is possible by examining the Wallner (stress) lines found on the edge of a fracture occurring due to tension as in Figure 14.1. The tension fractures occur as a glass sample is held firmly on all sides, such as in a window frame. First, radial fractures form radiating out from the point of impact. Second, concentric fractures occur connecting the radial fractures as in Figure 14.2. Wallner lines are not present on glass fragments broken due to heat or compression. Glass broken due to other manners (e.g., heat) does not exhibit Wallner lines.

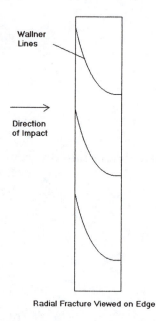

Radial Fracture Viewed on Edge

Figure 14.1. Glass fragment seen on the edge of a radial fracture showing Wallner lines and direction of impact. The Wallner lines form right angles on the reverse side of impact for radial fractures. However, the Wallner lines form right angles on the same side of impact for concentric fractures.

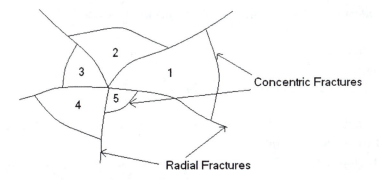

Figure 14.2. Fractured glass showing radial and concentric fractures. You will notice that fragments 1–5 all have two radial and one concentric fracture. This knowledge can be used to determine which is which for a glass fragment obtained at a crime scene (i.e., the two sides that have the Wallner lines in the same direction are radial fractures).

      Glass fragments can be distinguished from other items having the appearance of glass (e.g., plastics, minerals, salts) by determining its physical properties. Glass cannot be pierced with a pin, while most plastics can. Glass is not soluble in water, while many minerals and salts are. Glass is an amorphous solid and does not exhibit birefringence (two refractive indeces), while many plastics and minerals do. Birefringence can be determined by placing the sample on a microscope slide and viewing it under crossed-polars with a polarized light microscope. Samples that are birefringent will glow and colors are often seen.

      Density and refractive index are two common properties used for glass comparisons. Density of glass can be determined by measuring its mass and volume. Alternatively, relative density can be determined by placing the glass samples into a liquid of known density and observing their relative buoyancy. Glass samples of greater density than the liquid will sink while samples of lesser density will float. If the samples have the same density as the liquid, they will remain suspended.

      Refractive index can be determined by submerging the glass samples into a liquid of known refractive index and monitoring the movement of the Becke line (halo around the glass sample) as the objective of a microscope is focused away. The Becke line will move toward the medium (glass or liquid) of higher refractive index.

Purpose

To become familiar with glass fractures and Wallner (stress) lines indicating direction of impact. The student will also perform simple tests used to identify and compare glass samples.

Materials

Obtain the following materials:
1. One 10-mL graduated cylinder
2. One 250-mL beaker
3. One 4-mL glass test tube

The following materials and chemicals will be supplied as needed:
1. Spot plate
2. Unknown samples
3. Weigh boats
4. Fractured glass samples (heat and tension)
5. Cargille liquids
6. Float-sink method apparatus (bromoform/methanol mix in test tube)
7. Polarized light microscope
8. Analytical balance
9. Forceps
10. Microscope slides
11. Microscope cover slips

Procedures

Part A Glass Comparisons

1. Examine the broken glass samples prepared by the instructor. Determine how samples in group A and B were broken and, if possible, the direction of impact. **Record** your results.

2. Examine the glass samples prepared by the instructor placed in a float-sink apparatus in the fume hood. Determine if all of the samples have the same density by slowing heating of the apparatus in a water bath.* **Record** your results.

   * You do not need to overheat the apparatus. Gently heating for a few seconds should be adequate. Overheating will cause the lid to pop off. You may first try heating by simply holding it in your hands. Place the apparatus in a room-temperature water bath to cool in down for the next student.

Part B Glass Identification

1. Obtain three unknown powdered samples and determine which sample is glass. First, place a small amount of each in separate spot plate wells. Add a couple drops of water and **record** the water solubility of each. Glass is insoluble.

2. Next, place a small amount of each sample onto a microscope slide. You only need a tiny bit. You do **not** need to mount the sample with a cover slip. View each sample separately under the microscope and determine which samples are birefringent (the sample will appear to glow under the crossed polars of a polarized light microscope). A separate microscope slide is **not** required for each sample. **Record** your results.

Part C Determining Glass Density

1. Obtain a 4-mL glass test tube. Place a piece of paper over the test tube. Carefully strike the test tube with a heavy-solid object. ALWAYS WEAR YOUR SAFETY GOGGLES! Continue breaking the test tube until you have several very small pieces. You may be able to share fragments with other groups of students.

2. Obtain a 10-mL graduated cylinder. Find 15–20 pieces of glass that will fit into the mouth of the graduated cylinder. Obtain the mass of the fragments using an analytical balance and **record** your results.

3. Fill the graduated cylinder approximately half-full with water. **Record** the level of the water.

4. Slowly begin to add the weighed glass fragments into the top mouth of the graduated cylinder. Make sure that each glass fragment enters the water in the cylinder before adding another piece. Continue adding until all of the fragments are found in the graduated cylinder. **Record** the final level of the water.

5. Repeat steps 1–4 two more times using the same graduated cylinder.

Part D Determining Glass Refractive Index

1. Obtain a few very small (dust) fragments of glass broken from part C and place them onto a microscope slide. Immerse the sample in a Cargille liquid. Cover with a cover slip. **Record** the refractive index of the liquid used.

2. Examine the glass samples under the microscope and determine the relative refractive index of the glass with respect to the Cargille liquid. This can be done by monitoring the Becke line around the glass (Remember, the Becke line moves toward the substance of higher refractive index when focusing away). **Record** your results.

3. Repeat steps 1 and 2 until you find two Cargille liquids that bracket the refractive index of the glass. **Record** your results. A new glass sample may be required for each Cargille liquid; however, you can use the same microscope slide by cleaning it between tests. Proceed until the bracket is less than 0.05 RI units.

4. Clean up when finished.

Name:_____   Date:_____

Experiment 14 Worksheet

Results and Observations

Part A Glass Comparisons

- o How were the glass samples broken in group A (i.e., tension or heat)?

- o What allows you to determine this for group A samples?

- o Can you determine the direction of impact for group A samples? If so, what is the direction of impact?

- o How were the glass samples broken in group B (i.e., tension or heat)?

- o What allows you to determine this for group B samples?

- o Can you determine direction of impact for group B samples? If so, what is the direction of impact?

- o Do the submerged glass samples have comparable densities? Explain what evidence led you to your determination.

Part B Glass Identification

o   What unknown glass samples are water soluble?

o   What unknown glass samples are birefringent?

o   Based on your results, which sample is glass? Discuss your reasoning.

Part C Determining Glass Density

	Trial #1	Trial #2	Trial #3
Mass of Fragments			
Initial Volume			
Final Volume			
Density			

Average Density:_____

Part D Determining Glass Refractive Index

o   **Record** your observations for each Cargille liquid used. **Record** the refractive index of each Cargille liquid. Identify the two Cargille liquids with refractive indices that bracket the refractive index of the glass sample.

Experiment 14 Questions

1. In part C, if bubbles were introduced into the water as the glass samples were introduced, would the measured glass density be higher or lower than the actual? Explain.

2. Often, the technique used in part D is continued until a very narrow range of refractive indices is found bracketing the refractive index of the sample. Explain the importance of cleaning the glass slide between the changing of Cargille liquids if the same slide and glass sample are to be used for every analysis.

3. Based on your understanding of glass analyses, explain why glass is considered class evidence (i.e., only able to link to a class and not an exclusive origin).

4. It is known that refractive index and density are not completely independent of each other (i.e., as one increases so does the other). Explain what effect this has on the interpretation of your results.

Experiment 15
Fibers

Introduction

A fiber is the smallest component of a textile material. Many fibers can be woven into strands, yarn, rope, etc. Fibers, just like glass and hairs, are typically considered class evidence. However, as with all class evidence, the significance of the evidence is directly proportional to the number of class characteristics. A large number of correlating fibers of different class characteristics may potentially carry a case (e.g., Wayne Williams case).

Fiber analyses are performed by comparing the morphological features, the physical properties, and the chemical make-up of each questioned fiber to that of several known fibers. Although most analyses require the use of a microscope, many require the use of additional sophisticated instruments, such as a Fourier transform infrared spectrophotometer, a pyrolysis gas chromatograph, and/or a thin layer chromatograph.

Natural fibers (e.g., cotton, flax, and hemp) have unique microscopic features that allow most to be distinguished with the aid of a microscope. For example, an experienced fiber analyst can quickly identify cotton by viewing its flat and twisted ribbon appearance under the microscope. The polymer structure of natural fibers can also lead to various directions of twisting when dried. Often, fiber analysts will wet a fiber and dry it over a hot plate. Hemp and jute will display a counter-clockwise direction of twist when viewed on end, while flax will display a clockwise direction of twist.

Man-made fibers (e.g., nylon, polyester, and olefin) are typically less descriptive under the microscope when compared to natural fibers. Man-made fibers can be made to virtually any size, cross-section, hue, or appearance; thus, a microscopic analysis of a man-made fiber does not typically aid in identification (although it does aid in comparison). Other techniques for identification of man-made fibers include solubility tests and the use of instruments such as a Fourier transform infrared spectrophotometer (FTIR) or a pyrolysis gas chromatograph. The information acquired using these instruments is helpful in an identification since most man-made fibers have unique chemical compositions. However, these instruments are not helpful for the identification of a natural fiber since most natural fibers are composed of the same cellulose polymer.

Purpose

To introduce the student to many of the distinguishing properties of several natural and man-made fibers. Each student will become familiar with the drying-twist test, microscopic analysis, solubility testing, and FTIR analysis of fibers.

Materials

Obtain the following:
1. Natural fibers (cotton, hemp, jute, and flax)
2. Man-made fibers (polyester, acrylic, nylon, and olefin)
3. Seven microscope slides and covers
4. Spot plate

5. Stereoscopic microscope
6. Forceps

The following materials will be supplied as needed:
1. Compound/biological microscope
2. Xylene
3. Permount^{TM}
4. Dry-twist test apparatus (beaker, water, and a hot plate)
5. Solubility test chemicals (HFIP, formic acid, dimethyl formamide, glacial acetic acid, acetonitrile, nitric acid, 75% sulfuric acid)
6. Fourier transform infrared spectrophotometer (FTIR)

Procedures

Part A Microscopic Analysis

1. Each group prepare Permount^{TM} microscope slides of cotton, hemp, flax, jute, silk, nylon, and polyester. Do this by placing each fiber on a separate microscope slide, adding a drop or two of Permount^{TM}, and placing a microscope slide on top.

2. Using the technique demonstrated by the instructor, prepare cross-sectional preparations of both nylon and polyester fibers.

3. View each of the above slides under the compound/biological microscope. **Record** your observations.

Part B Dry-Twist Test

1. Obtain one jute fiber. Using forceps, wet the fiber in a beaker of water. While holding the fiber with forceps over a hot plate, view the fiber end-on and **record** the direction of twist.

2. Repeat step 1 for hemp.

Part C Solubility Tests

1. Perform solubility tests. Obtain several polyester fibers. Place the fibers into the well of a spot plate. While viewing under a stereoscopic microscope in a fume hood, add the first solvent in the flow chart in Figure 15.1 (i.e., formic acid). **Record** your results. Continue adding and removing chemicals as suggested by the flow chart. **Do not mix chemicals. Always use a separate well with a new fiber for each test.**

2. Repeat step 1 for nylon.

3. Test the solubility of silk in hydrochloric acid, formic acid, nitric acid, and sulfuric acid. **Record** your results.

Part D Instrumental Analysis

1. Your instructor will demonstrate how to use the FTIR for the identification of unknown fibers. **Record** all procedures.

2. Use the FTIR to determine the type of two man-made fibers.

Part E Unknown Fibers

1. Identify two unknown fibers using whatever techniques you choose.

2. Clean up when finished.

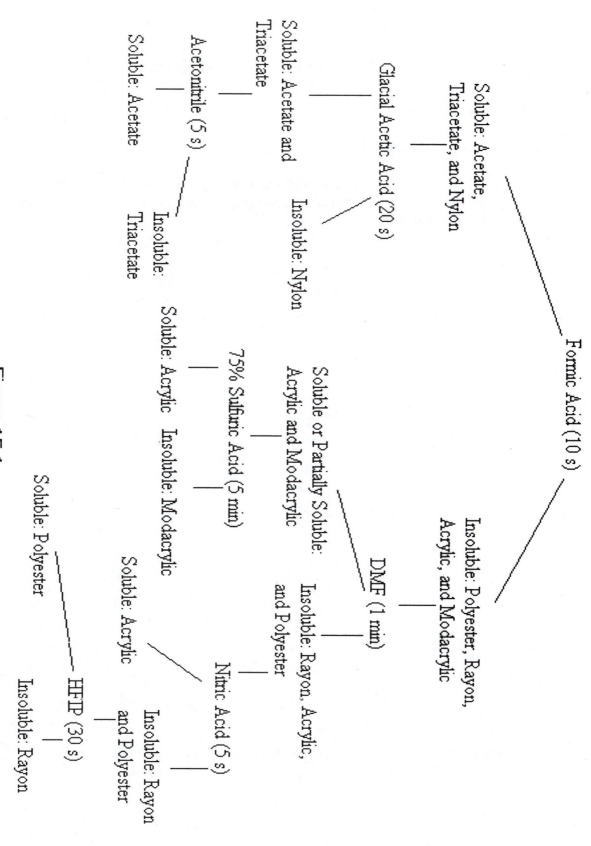

Figure 15.1

Experiment 15 Worksheet

Results and Observations

Part A Microscopic Analysis

- In the space provided, **record** the macroscopic and microscopic appearance (using a compound/biological microscope) of hemp, jute, flax, cotton, nylon, polyester, and silk.

- Describe the cross-section of the prepared nylon and polyester fibers.

Part B Dry-Twist Test

o **Record** the observed direction of twist for hemp and jute.

o Describe any difficulties encountered when performing the dry twist test.

Part C Solubility Tests

o **Record** your observations for each solvent used to determine the solubility of polyester (following the flow chart).

o **Record** your observations for each solvent used to determine the solubility of nylon (following the flow chart).

o **Record** the observed solubility of silk in nitric acid, formic acid, hydrochloric acid, and sulfuric acid.

Part D Instrumental Analysis

o **Record** all procedures necessary for the operation of the FTIR.

o Print out the FTIR acquired of a selected synthetic fiber. **Record** the identification of the fibers.

Part E Unknown Fibers

o Describe in detail the results of the techniques used for the identification of both unknown fibers.

o Explain why the above techniques were chosen for the unknown fiber identification.

o What is the identification of the two unknown fibers?

Experiment 15 Questions

1. Discuss the advantages and limitations of using fiber cross-sections for both comparison and identification of natural and man-made fibers.

2. Discuss the advantages and disadvantages associated with the solubility testing of fibers.

3. Discuss how each fiber in a blend of fibers (that are difficult to physically separate) might be determined using the analysis techniques available in this lab.

Experiment 16
UV-vis Analysis of Food Dyes

Introduction

Many food products are colored with common food dyes such as yellow #5, yellow #6, red #3, blue #2, red #40, and green #3. All of these dyes are colorful compounds composed of conjugated double bonds. Compounds exhibit color when capable of absorbing various wavelengths of light within the visible region of the electromagnetic spectrum. The color observed is typically the complementary color to the color of light absorbed. For example, a blue compound appears blue because it is capable of absorbing red light.

Absorption of light occurs when an electron is excited from a lower energy level to a higher energy level. The conjugation of double bonds (alternating single and double bonds along the length of a molecule) lowers the energy of transition such that light may be absorbed in the visible region of the electromagnetic spectrum; otherwise, absorption might occur in the ultraviolet region (higher energy) and the compound would not appear colored. In addition to conjugated double bonds, many colorful compounds also contain doubly bonded nitrogen atoms (-N=N-) called azo groups.

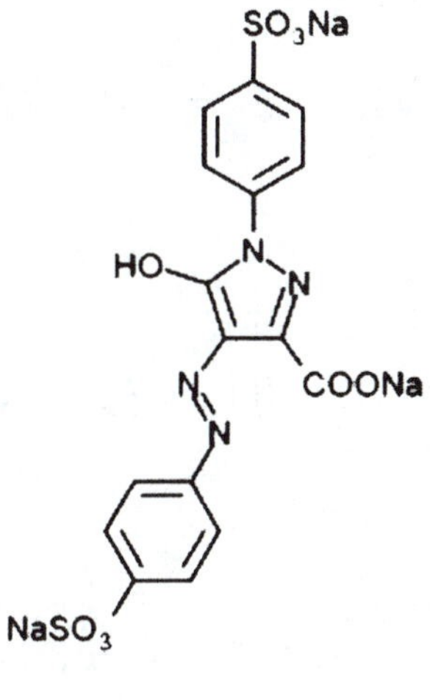

Figure 16.1. Yellow #5, tartrazine, an azo compound with excessive double-bond conjugation.

Yellow #5 is a very common food dye found in products such as gelatin, pudding, macaroni and cheese, soft drinks, and pickle juice. This product is moderately flammable and a very mild skin, eye, and digestive tract irritant. In mice, the lethal dose in 50% of the population (LD50) is the astronomically high amount of 12750 mg/kg. For comparison, most food products can be classified as mild irritants. Sodium chloride (table salt) is a mild irritant with an LD50 of 3000 mg/kg in rats (one-fourth that of tartrazine in mice!). However, as with most food dyes, it is not wise to include more than what is necessary for a particular product.

Ultraviolet/visible light spectrophotometry (UV-vis) is an analytical technique used to measure the extent of visible and ultraviolet light absorption at determined wavelengths, or across a range of wavelengths. When measuring the absorbance across a range of wavelengths, the instrument is capable of generating spectra (graphs of absorbance versus wavelength) representative of the samples. Because these instruments measure both visible and ultraviolet light absorbance, spectra include more than just information associated with the color of the samples.

UV-vis will be used in this experiment to measure the amount of yellow #5 in Mountain Dew® and Sam's Choice Mountain Lightning®, both having a similar appearance and taste (debatable). In order to do this, a plot must first be generated characterizing the extent of absorbance occurring due to the presence of yellow #5 and its concentration. These plots are called calibration curves and are prepared by measuring the absorbance occurring for solutions of known yellow #5 concentrations. After preparing the calibration curve, the absorbance of each soft drink can be measured and the curve used to determine yellow #5 concentration.

Purpose

To introduce the student to UV-vis instrumentation, food dyes, and calibration curves. Each student will prepare their own calibration curve for yellow #5 and determine the concentration of yellow #5 of two soft drinks.

Materials

Obtain the following materials:
1. Two 50-mL beakers
2. One 250-mL beaker

The following materials and chemicals will be supplied as needed:
1. Five 100-mL volumetric flasks
2. Tartrazine (yellow #5)
3. Distilled water
4. 50-, 25-, 10-, and 5-mL pipettes
5. UV-vis spectrophotometer
6. UV-vis cuvettes
7. Mountain Dew®
8. Sam's Choice Mountain Lightning®
9. Sonicator (optional, to de-carbonate the soft drinks)

Procedures

Part A Preparation of Standards and Samples

1. Prior to lab, a 0.000100 M solution of tartrazine was prepared. This will be your stock solution. Obtain approximately 100 mL of this solution in a 250-mL beaker and take it to your workstation.

2. Using a 50-mL pipette, transfer 50 mL of the stock tartrazine solution to a 100-mL volumentric flask.* Fill to the 100-mL mark with distilled water and label as solution A. Calculate and **record** the concentration of solution A.

   * If you have never used a pipette, ask you instructor to demonstrate.

3. Repeat step 2 using a 25-, 10-, and 5-mL pipettes, preparing three more solutions of diluted tartrazine stock solution in three separate 100-mL volumetric flasks. Label each B, C, and D, respectively. Calculate and **record** their concentrations.

4. Obtain approximately 25 mL of Mountain Dew® and 25 mL of Sam's Choice Mountain Lightning ® in two separate 50-mL beakers. If the soft drink samples are still carbonated, you may need to briefly place each into a sonicator bath to remove carbon dioxide bubbles. **Record** their appearance.

Part B Measurement of Absorbances

1. Your instructor will demonstrate how to use your UV-vis spectrophotometer. **Record** all of the necessary steps for its operation.

2. Measure and **record** the absorbance of each solution (all standards and both samples) at 427 nm and 335 nm.

3. Clean up when finished.

Experiment 16 Worksheet

Results and Observations

Part A Preparation of Standards and Samples

Solution	mL of Stock Solution	Total Volume	Concentration
A			
B			
C			
D			

- o **Record** the appearance of both the Mountain Dew® and Sam's Choice Mountain Lightning samples.

Part B Measurement of Absorbances

o   **Record** all of the necessary steps for the operation of your UV-vis spectrophotometer.

Solution	Absorbance, 427 nm	Absorbance, 335 nm
A		
B		
C		
D		
Mountain Dew®		
Sam's Choice Mt. Lightning®		

Experiment 16 Questions

1. Prepare a calibration curve for absorbance data at both 427 and 335 nm (you may choose to use Excel or graph paper). Which wavelength produces the *best* calibration curve? Hint: It will be the one that shows the greatest change in absorbance with change in concentration. Attach your curves to your worksheet.

2. Using the above-prepared *best* calibration curve and the absorbance measurements for Mountain Dew® and Sam's Choice Mountain Lightning®, determine the concentration of tartrazine in each soft drink. (If you prepared your curves in Excel, you may use the generated line equation; otherwise, draw your own best-fit line through your data points on the graph paper to obtain the answers.)

3. Based on your answer to question 2, is there a significant difference between the amount of tartrazine added to Mountain Dew® versus Sam's Choice Mountain Lightning®? If you were working in a crime laboratory, explain whether or not you would be able to distinguish Mountain Dew® from Sam's Choice Mountain Lightning® solely based on tartrazine concentration. (Consult your peers to see if they obtained similar results and include their findings in your answer.)

4. When comparing two substances of similar appearance using UV-vis spectrophotometry, often more than one wavelength is selected. The wavelength chosen in question #1 is assumed to be the wavelength of maximum absorbance for tartrazine. However, many compounds may have more than one wavelength of absorbance. In addition, soft drinks such as Mountain Dew® and Sam's Choice Mountain Lightning® contain more than just tartrazine. These other compounds can also absorb UV and visible light. In this experiment, a second wavelength was chosen. Explain if, and how, this additional wavelength can be used to further distinguish Mountain Dew® from Sam's Choice Mountain Lightning®.

5. One can of soda pop is 355 ml. Determine the number of milligrams of tartrazine that are consumed when drinking one can of Sam's Choice Mountain Lightning® (the molecular weight of tartrazine is 534.37 grams/mole). If the LD50 of tartrazine in humans is assumed to be 12750 mg/kg (as it is for mice) and the average human weighs 70 kg, how many cans of Sam's Choice Mountain Lightning® would an individual have to drink to die from a tartrazine overdose? Is this possible?

Experiment 17
Gunshot Residue

Introduction

If a gun is fired at close range, gunshot residue will typically accumulate on the target surrounding the bullet hole. Gunshot residue may be composed of large particles of burned and unburned propellant, fine burned particles, particles of metal from the bullet, and/or microscopic particles of the primer. Based on the area covered by the gunshot residue powder pattern, and the types and amount of residue present (i.e. soot and/or stippling), one may be able to estimate the distance from which a gun was fired up to about 2 feet. After this distance, the debris becomes too sparse.

Debris surrounding an assumed bullet hole may be of a nature foreign to gunshot residue, i.e., a bullet did not create the hole, and/or the target may have characteristics (e.g., dark colored) that are not conducive to the visual examination of the residue powder pattern. To aid analyses under these circumstances, presumptive color tests for gunshot residue can be used to both enhance the powder pattern and identify components common to gunshot residue.

The three common color tests employed are: 1) diphenylamine test, 2) Griess test, and 3) rhodizonic acid test. The diphenylamine test is a test for the presence of nitrates, nitrites, and other oxidizing reagents. A positive test results in a blue color that may fade to a yellowish-olive green color dependent on the solution matrix and the concentration of the sample. The Griess test is a test for solely nitrites generated during the combustion of gun powders. A positive test results in a rosy-orange or reddish color also dependent on the solution matrix and the concentration of the sample. The rhodizonic acid test is for the presence of lead and, to a lesser extent, other heavy metals (e.g., barium). A positive test for lead results in a scarlet red color that changes to a violet color with the addition of hydrochloric acid. This change in color distinguishes a positive lead test from other heavy metals.

When performing the color tests in this experiment, it is best to run a control (blank) right next to the well containing the sample being tested. A control contains all of the chemicals needed for the reaction less the sample. This may aid you in determining a color change.

Purpose

To examine how three common color tests (i.e., Griess test, sodium rhodizonate test, and diphenylamine test) can presumptively identify and enhance the presence of gunshot residue. Each student will perform the Griess test and the diphenylamine test on both burned and unburned black and smokeless powder.

Materials

The following materials and chemicals will be supplied as needed:
1. Diphenylamine
2. Sulfuric acid

3. Sodium rhodizonate or sodium rhodizonic acid
4. Acetic acid
5. Methanol
6. Hydrochloric acid solution
7. Sulfanilic acid
8. 1-naphthylamine
9. Black powder
10. Smokeless powder
11. Watch glass
12. Spot plate
13. Spatula
14. Elemental lead

Procedures

Part A Prepare Color Test Solutions

1. Prepare 2 mL of diphenylamine solution by weighing out 0.2 g diphenylamine and placing it into a test tube. Next, add 2 mL of sulfuric acid to this test tube.

2. Prepare a solution of rhodizonic acid by weighing out 0.004 g of rhodizonic acid and placing it into a test tube. Next, add 2 mL of distilled water and five drops of acetic acid to the rhodizonic acid.

3. Prepare a solution of 1-naphthylamine by weighing out 0.01 g of 1-naphthylamine and placing it into a test tube. Next, add 2 mL of methanol to the test tube.

4. Prepare a solution of sulfanilic acid by weighing out 0.02 g of sulfanilic acid and placing it into a test tube. Next, add approximately 1.4 mL of distilled water and 0.6 mL of acetic acid to the test tube. The last two solutions are used together in the Griess test for nitrites.

Part B Unburned Gunpowder Samples

1. Obtain a very small amount of black powder and place equal amounts into four separate wells of a spot plate. Each well should just contain 1–4 flakes of powder. Add 2–3 drops of water to two of the wells.

2. Add one drop of diphenylamine solution to one well containing dry powder and one well containing the powder with water. This is a test for the presence of nitrates, nitrites, and other oxidizers. **Record** both immediate and subsequent results.

3. Next, add one drop of 1-naphthylamine solution and then one drop of sulfanilic acid solution to both a separate dry sample of powder and a separate wet sample

of powder. This is a test for nitrites. **Do not mix chemicals from this step with the chemicals in step 2. Record** both immediate and subsequent results.

4. Repeat steps 1–3 for smokeless powder.

Part C Burned Gunpowder Samples

1. Obtain a small amount of black powder and place it in a watch glass. Place the watch glass in a fume hood and light the powder.

2. Using a spatula, scrape off some of the residue into two wells of a spot plate.

3. Next, add a few drops of water to the residue left in the watch glass and swirl it around. Be careful not to spill. Remove the water found in the residue and place equal amounts into two separate wells of a spot plate.

4. Add a drop of diphenylamine solution to one well prepared in step 3 and one drop of both 1-naphthylamine solution and sulfanilic acid solution to the second well. **Record** both immediate and subsequent results.

5. Perform step 4 on the dry samples scraped from the watch glass.

6. Repeat steps 1–5 for smokeless powder.

Part D Lead Analysis

1. Obtain a piece of paper, a pencil, and a lead bullet. Make markings on the piece of paper with both the pencil and the lead bullet.

2. Add one drop of rhodizonic acid solution to the markings on the paper. **Record** both immediate and subsequent results.

3. Add one drop of hydrochloric acid solution to the markings developed with the rhodizonic acid solution. **Record** both immediate and subsequent results.

4. Clean up when finished.

Experiment 17 Worksheet

Results and Observations

Part B Unburned Gunpowder Samples

- o Observations of dry black powder with diphenylamine solution

- o Observations of wet black powder with diphenylamine solution

- o Observations of dry black powder with Griess test solutions (i.e., 1-naphthylamine and sulfanilic acid)

- o Observations of wet black powder with Griess test solutions (i.e., 1-naphthylamine and sulfanilic acid)

- o Observations of dry smokeless powder with diphenylamine solution

- o Observations of wet smokeless powder with diphenylamine solution o Observations of dry smokeless powder with Griess test solutions (i.e., 1-naphthylamine and sulfanilic acid)

o Observations of wet smokeless powder with Griess test solutions (i.e., 1-naphthylamine and sulfanilic acid)

Part C Burned Gunpowder Samples

o Observations of dry burned black powder with diphenylamine solution

o Observations of wet burned black powder with diphenylamine solution

o Observations of dry burned black powder with Griess test solutions (i.e., 1-naphthylamine and sulfanilic acid)

o Observations of wet burned black powder with Griess test solutions (i.e., 1-naphthylamine and sulfanilic acid)

o Observations of dry burned smokeless powder with diphenylamine solution

o Observations of wet burned smokeless powder with diphenylamine solution

o Observations of dry burned smokeless powder with Griess test solutions (i.e., 1-naphthylamine and sulfanilic acid)

o Observations of wet burned smokeless powder with Griess test solutions (i.e., 1-naphthylamine and sulfanilic acid)

Part D Lead Analysis

o Observations of graphite markings with rhodizonic acid solution and hydrochloric acid

o Observations of lead markings with rhodizonic acid solution and hydrochloric acid

Experiment 17 Questions

1. Were nitrates or nitrites present in the unburned black powder samples? Explain.

2. Were nitrates or nitrites present in the burned black powder samples? Explain.

3. Were nitrates or nitrites present in the unburned smokeless powder samples? Explain.

4. Were nitrates or nitrites present in the burned smokeless powder samples? Explain.

5. Which test, wet or dry (if any), gave more easily interpretable results using the diphenylamine solution on unburned black powder? Why might this be the case (Hint: solubility rules)?

6. How did the unburned wet smokeless powder diphenylamine test results compare to the wet burned smokeless powder diphenylamine test results? Did both tests give positive results? If not, or if yes, what is your explanation based on your results?

7. Based on your results, are pencils made of lead? Explain.

Experiment 18
Introduction to Chemical Instrumentation

Introduction

The separation of compounds is often required prior to compound identification (e.g., heroin must be separated from its diluent(s) prior to identification). This is typically accomplished by employing chromatographic or electrophoretic separation techniques. The following experiment will focus on the former and introduce the student to gas chromatography.

Gas chromatography (GC) typically employs a long (e.g., 30 m) fused-silica capillary as a separation column. A stationary phase is coated on the inside of the capillary, and a gas (i.e., helium or hydrogen) serving as a mobile phase carries the compounds through the column. Different from thin-layer chromatography, GC detects compounds as they elute (come off) from the end of the column. The elution or retention time is used to presumptively identify the compound.

A wide variety of detectors are available to detect compounds after separation. Some detectors simply indicate the presence of a compound while others offer additional qualitative information used for compound identification. The ultraviolet/visible light (UV-vis) detector is often used to simply determine the retention time of a compound. The Fourier transform infrared spectrophotometer (FTIR) and the mass spectrometer (MS) are commonly used in crime laboratories for compound identification.

The UV-vis detector measures the amount of visible and ultraviolet light absorbed by a sample. It can employ a single wavelength to determine the retention time of a compound and its concentration, or it can use multiple wavelengths to presumptively identify a compound. It will be used in this experiment to presumptively identify a pure compound that needs no separation.

The FTIR detector generates a "fingerprint," or a unique spectrum of infrared light absorbance, that can be used for compound identification. All spectra generated represent compound structure as a function of infrared light interaction. The wavelengths of infrared light absorbed in conjunction with their relative intensities serve as a means of individualization. Spectra are very compound-specific and are matched to spectra stored in libraries for compound identification. It will be used in this experiment to identify a pure compound that needs no separation.

The MS detector is also capable of generating a compound "fingerprint," or spectrum; however, it is in a completely different manner from that used for FTIR. The spectra generated are a function of compound mass rather than infrared light interaction. A compound is bombarded with an electron beam, breaking it into fragments. The mass and relative intensities of the fragments serve as a means of individualization. MS will be used in this experiment to detect and identify a compound separated using gas chromatography.

Purpose

To introduce the student to chemical instrumentation commonly used in the crime laboratory. The student will become familiar with paper filtration, gas chromatography mass spectrometry, infrared spectrophotometry, and ultraviolet light spectrophotometry. The potential for identification using each instrument will also be presented.

Materials

Obtain the following materials:
1. Three 50-mL beakers
3. One 25-mL graduated cylinder
4. One mortar and pestle
5. One pill
6. One spatula

The following materials and chemicals will be supplied as needed:
1. Methanol
2. GC/MS instrument and GC vials
3. UV-vis instrument cuvettes
4. FTIR instrument
5. Filter paper #4

Procedures

Part A Prepare Pill Powder

1. Clean and dry a mortar and pestle using distilled water.

2. Grind a pill into a fine powder using the mortar and pestle. If the pill is coated, break the capsule open by hitting it several times with the pestle. You should be able to remove the gel coating before grinding. **Record** the appearance of the pill sample before and after grinding.

3. Clean and dry three 50-mL beakers.

4. Measure and **record** the mass of one 50-mL beaker. Using a spatula, slowly add approximately 0.1 grams of pill powder to the beaker on the balance. **Record** the final mass. DO NOT dispose of the extra pill powder; you will need it in part E.

5. Under the fume hood, add approximately 20 mL of methanol (using a 25-mL graduated cylinder) to the beaker containing the pill powder.

6. Carefully swirl the beaker containing the methanol and pill powder so as to dissolve as much as possible. **Record** the appearance of the solution.

Part B Filter Pill Sample and Prepare UV-vis Sample

1. Obtain two pieces of filter paper.

2. Fold the filter paper circle in half twice, making four equal sections. Hold the folded paper with the point down and the curve up. Find the flap on the extreme right or left and open the cone, making a funnel.

3. Place the funnel prepared in step 2 in a clean 50-mL beaker and carefully pour the methanol/pill powder solution into the funnel. You may have to pour small amounts at a time and occasionally raise the funnel.

4. After filtering, check the filtrate (solution poured into the second beaker through the funnel) to make sure that it is completely clear. **Record** its appearance. If it is not clear, run it through a second filter paper prepared as in step 2 and **record** its appearance.

5. Prepare another funnel and, using a 25-mL graduated cylinder, filter 20 mL of pure methanol into a third 50-mL beaker. **Record** the appearance of the filtered methanol solution.

Part C Analyze Pill Sample Using UV-vis

1. Your instructor will demonstrate how to use the UV-vis instrument. **Record** all steps necessary for its operation.

2. Pour a small amount of the filtered methanol into a cuvette as demonstrated by the instructor.

3. Pour a small amount of filtered methanol/pill solution into a cuvette as demonstrated by the instructor.

4. Place both into their appropriate location in the UV-vis instrument and acquire a spectrum for the pill sample as demonstrated by the instructor.

5. Print out the spectrum.

6. Dispose of solutions **in the cuvettes** in an appropriate waste container. Dispose of the filtered methanol solution in the same waste container. DO NOT DISPOSE OF THE FILTERED METHANOL/PILL SOLUTION. YOU WILL NEED IT IN PART D.

Part D Prepare GC/MS Sample and Analyze

1. Pour 1.0 mL of filtered methanol/pill solution into a 25-mL graduated cylinder. In the fume hood, fill the graduated cylinder to the 10-mL mark with methanol. This is done to dilute the sample to an appropriate concentration for GC/MS analysis.

2. Mix the solution well by gently tapping the bottom of the graduated cylinder.

3. Obtain a GC vial from the instructor and fill it about ¾ full with the solution prepared.

4. Your instructor will demonstrate how to use the GC/MS instrument. **Record** all steps necessary for its operation.

5. Place the vial in the GC autosampler and analyze as demonstrated by the instructor. The instructor may also suggest performing a manual injection.

6. Search the library to identify the pill peak as demonstrated by the instructor. **Record** the results.

7. Print the chromatogram and the mass spectrum for the pill sample.

Part E Analyze Pill Sample Using Fourier Transform Infrared Spectrophotometry (FTIR)

1. Your instructor will demonstrate how to use the FTIR instrument. **Record** all steps necessary for its operation.

2. Place a **very** small amount of pill powder on the crystal of the attenuated total reflectance accessory for FTIR.

3. Analyze the sample as demonstrated by the instructor.

4. Search the library to identify the pill sample and print the infrared spectrum of the sample with its identification.

5. Clean up when finished.

Name:_____   Date:_____

Experiment 18 Worksheet

Results and Observations

Part A Prepare Pill Powder

Record the appearance of pill sample before grinding.

Record the appearance of pill sample after grinding.

Measuring approximately 0.1g of pill powder

---	Beaker	Beaker + Sample	Sample
Mass, g			

Record the appearance of unfiltered methanol/pill powder solution.

Part B Filter Pill Sample and Prepare UV-vis Sample

o   Appearance of filtered methanol/pill solution (First trial) o   Appearance of filtered methanol/pill solution (Second trial, if necessary)

o   Appearance of filtered methanol solution

Part C Analyze Pill Sample Using UV-vis

o   **Record** all steps necessary for the operation of the UV-vis instrument.

Part D Prepare GC/MS Sample and Analyze

o   **Record** all steps necessary for the operation of the GC/MS instrument.

o   What is the retention time of the pill peak?

○ What did the library identify this peak to be?

Part E Analyze Pill Sample Using Fourier Transform Infrared Spectrophotometry

○ **Record** all steps necessary for the operation of the FTIR instrument.

○ What do you see under the microscope when viewing the pill powder sample? Be descriptive.

○ What did the library identify the powder to be?

Experiment 18 Questions

1. If the UV-vis instrument measures the absorbance of light occurring as the light passes through the cuvette containing the sample dissolved in methanol, why was it important to prepare a "blank" methanol sample when analyzing the pill sample using UV-vis spectroscopy?

2. How would the results have differed **for all three analyses** if a gel coating would have been ground with the pill?

3. Which two tests were confirmatory and capable of generating a compound "fingerprint"? Explain.

4. Briefly explain what information is being acquired to identify the pill in each analysis (i.e., GC/MS, UV-vis, and FTIR).

Experiment 19
Microscopy

Introduction

The microscope is one of the most valuable and versatile tools available to the forensic scientist. Microscopes are commonly used for the analysis of hair, fibers, drugs, fingerprints, soil, glass, biological fluids, tool marks, physical matches, bullets, and cartridge casings. Despite the development of highly sophisticated instrumentation over the last century, the microscope maintains its importance in the crime laboratory.

There are many types of microscopes, i.e., simple (magnifying glass) compound/biological, stereoscopic, polarized light, comparison, microspectrophotometer, and electron microscope. The forensic scientist regularly uses most of these microscopes. This experiment will introduce the student to the stereoscopic and the compound/ biological microscope which are found in virtually all crime laboratories.

The capabilities of a microscope go beyond simple magnification of an object. The microspectrophotometer, polarized light, and electron microscopes are routinely used to additionally acquire information associated with the chemical makeup of the sample being analyzed. Such information is routinely used for the purpose of comparison and/or identification.

Purpose

To introduce the student to the microscope. Students will become familiar with the components and the operation of both a stereoscopic and compound/biological microscope.

Materials

Obtain the following materials:
1. Pencil or pen
2. **Bring a picture(s) showing the components of a stereoscopic and a compound microscope to lab. You may need to reference a textbook or surf the Web.**

The following materials and chemicals will be supplied as needed:
1. Stereoscopic microscope
2. Compound/biological microscope
3. Microscope slides
4. Illuminator
5. Sample hair slides

Procedures

Part A Stereoscopic Microscope Introduction

1. Obtain a stereoscopic microscope and bring it to your workstation. You may need to work in groups and/or go onto part C if there are not enough stereoscopic microscopes.

2. Examine the lens to make sure that they are free of dust, dirt, and grease. If they are dirty, you may ask for lens paper (if available) and clean the lenses by lightly wiping them with the dry lens paper. Never use anything else to clean the lenses. Do not clean the lenses if it is not necessary! Excessive wiping my damage the lenses.

3. Using your picture, review the components of the microscope. Make sure that you can identify the oculars, objectives, body tubes, stage, illuminator(s), base, arm, coarse and fine focus, and zoom (if applicable). Draw a picture of your microscope and label all parts.

Part B Stereoscopic Microscope Operation

1. Place a pencil or pen on the stage of the microscope. Make sure that the sample is adequately illuminated with an epi (reflected) illumination source.

2. Start your examination at the lowest magnification; you have greater field of view and depth of focus. **Record** your observations and total magnification.*

    * The total magnification is equal to the magnification of the ocular times that of the objective. A zoom focus may require the reading of a dial having the magnification of the objective.

3. Focus on the very tip of the pencil and **record** your observations. If the microscope is binocular, you may want to adjust the distance between the oculars to comfortably fit your eyes. Also, if the microscope has additional focusing on one of the oculars, you may want to focus each eye separately. Start by closing the eye found over the ocular that cannot be focused independently. Focus on the specimen. Open the second eye and close the first. Manually adjust the single ocular focus for your individual eye such that both oculars are in focus when viewing with both eyes.

4. Focus on the tip of the pencil at an increased magnification. **Record** your observations and total magnification. Make sure to discuss depth of focus and field of view differences.

5. Repeat steps 3 and 4 by examining the letters found on this page.

Part C Compound/Biological Microscope Introduction

1. Obtain a compound/biological microscope and bring it to your workstation. You may need to work in groups if there are not enough of these microscopes.

2. Examine the lenses to make sure that they are free of dust, dirt, and grease. If they are dirty, you may ask for lens paper (if available) and clean the lenses by lightly wiping them with the dry lens paper. Never use anything else to clean the lenses. Do not perform this step if cleaning is not necessary.

3. Using your picture, review the components of the microscope. Make sure that you can identify the oculars, objectives, body tube, stage, illuminator, base, arm, coarse and fine focus, turret, condenser, and the field (illuminator) and aperture stops. Draw a picture of your microscope and label all parts. Determine and **record** whether the stage or objectives move up and down when adjusting the focus. Also determine and **record** the direction needed to turn the focus (clockwise or counterclockwise) in order to increase the distance between the objective and the stage.

Part D Compound/Biological Microscope Setup (Köhler Illumination*)

* Köhler illumination offers uniform brightness to an object, minimizes distortion, and allows for the best focus possible. Setting up Köhler illumination is a way of "tuning" the microscope. You will become familiar with the basics in the steps that follow by adjusting the illuminator, aperture stop, field stop, condenser, and objectives where applicable. If one of the components cannot be adjusted on your microscope, you may skip this step; however, it is recommended that you understand what would have been done.

1. Place the 10× objective into place. Adjust its height to 2–3 mm above the stage.

2. Turn on the illuminator. If it has an adjustable intensity, make it as bright as possible while still allowing for comfortable viewing.

3. Place a specimen slide on the stage of the microscope.

4. For binocular microscopes, adjust the ocular separation to fit your eyes. Also, if the microscope has additional focusing on one of the oculars, you may want to focus each eye separately. Start by closing the eye found over the ocular that cannot be focused independently. Focus on the specimen. Open the second eye and close the first. Manually adjust the single ocular focus for your individual eye such that both oculars are in focus when viewing with both eyes.

5. Open both the field and aperture stops completely. You should have a full field of view with a blurry specimen. **Record** your observation.

6. Bring the specimen into focus by increasing the distance between the specimen and objective, first with the coarse and then fine adjustments. It is always best to start with the objective close to the specimen and then to focus away. It is difficult to know the distance between the specimen and the objective when viewing through the oculars, and a specimen may be crushed when focusing toward the stage.

7. Completely close the field stop so that all you can see is a fuzzy (or multi-colored) circle of light around your specimen. Make sure that you can at least see part of your specimen. **Record** your observations.

8. Focus the edge of the circle (i.e., the diaphragm edge) by moving the condenser up or down. The circle of light should now appear as a polygon. **Record** your observations.

9. If the circle of light is not in the center of the blackened area, adjust the condenser using the screws on the side.

10. Open the field stop until the lighted area just fills the circular field. The polygon should be just touching the outer ridge of the circular field.

11. Remove one of the oculars. Peer down the body tube and view the lamp filament. **Record** your observations.

12. Adjust the lamp filament into focus by moving the lamp with adjusting screws.

13. Open or close the aperture stop until the polygon fills about 70–80% of the field of view.

14. Replace the ocular. You now have Köhler illumination. If the objectives are all aligned (we will assume that they are), you will not need to repeat the process for each objective used. Otherwise, there would be an additional step in the above procedure to align the objectives.

Part E Compound/Biological Microscope Operation

1. Place a hair slide onto the stage of the microscope; find the tip of the hair at lowest magnification. **Record** your observations and total magnification.

2. Increase the magnification to the highest magnification (do not use any objectives requiring oil immersion). You may need to slightly refocus. **Record** your observations and total magnification.

3. Find the root of the hair at lowest magnification and **record** your observations and total magnification.

4. Increase the magnification to the highest magnification not requiring oil immersion and **record** your observations and total magnification.

5. Focus on a portion of the shaft (mid-length) of the hair at lowest magnification. **Record** your observations and total magnification.

6. Increase the magnification to the highest magnification and **record** your observations and total magnification.

7. Clean up when finished.

Name:_____   Date:_____

Experiment 19 Worksheet

Results and Observations

Part A Stereoscopic Microscope Introduction

Draw a picture of the stereoscopic microscope and label its parts.

Part B Stereoscopic Microscope Operation

Observations of pencil at low magnification

Observations of pencil at high magnification

Observations of letters at low magnification

Observations of letters at high magnification

Part C Compound/Biological Microscope Introduction

o   Draw a picture of the compound/biological microscope and label its parts. Also, **record** all information requested in the procedures of this part.

Part D Compound/Biological Microscope Setup (Köhler Illumination)

Step 5 observations

Step 7 observations

Step 8 observations

Step 11 observations

Part E Compound/Biological Microscope Operation (Be very descriptive)

Observations of hair tip at low magnification

Observations of hair tip at high magnification

Observations of hair root at low magnification

Observations of hair root at high magnification

Observations of hair shaft at low magnification

Observations of hair shaft at high magnification

Experiment 19 Questions

1. Based on your understanding of the operation and capabilities of a stereoscopic and a compound/biological microscope, which would be best to view a clear river water sample and why?

2. Discuss three differences between the stereoscopic and the compound/biological microscopes.

Experiment 20
Footwear Impressions and Physical Matches

Introduction

Footwear impressions are pattern transfers occurring when the pattern of the sole of an individual's footwear is left behind. Often, footwear impressions are *not* exploited to their fullest potential. This may be because many believe footwear impressions lack evidential value due to the large number of similar shoes sold. Also, investigators simply may not know where to look and/or potentially trample evidence. However, footwear impressions **do** have the potential to be as individual as fingerprints. In addition, such evidence often offers insight into events that occurred during the commission of the crime.

A physical match, or a jigsaw fit, can be described as two separated items coming together with the pattern of one mirroring that of the other (e.g., two puzzle pieces fitting together). Such a match is highly desirable and virtually conclusive as to the commonality of the two items' origin. Physical matches can occur with all types of evidence (e.g., paper, sticks, fabric, glass, or virtually anything broken ripped or torn). One must be cautious when attempting a physical match such that one does not grind two items together forcing them to match.

Purpose

To become familiar with footwear impression comparisons and physical matches. Each student will perform tasks allowing him or her to recognize the significance of footwear impression comparisons and physical matches.

Materials

Obtain the following materials:
1. Seven pieces of white paper (supply your own)
2. One pair of shoes (supply your own)

The following materials and chemicals will be supplied as needed:
1. Dirt or sand
2. Scotch tape
3. Comparison prints
4. Pan for 3-D footwear impression
5. Broken wooden cotton swabs
6. Plaster-of-Paris
7. Mixing spatula

Procedures

Part A Paper Footwear Impression

1. Choose one of your shoes to be examined. **Do not** remove the shoe. Acquire some fine dry dirt and place a small amount on the floor (not a lot of dirt is needed; keep the custodians happy).

2. While wearing your shoe, gently rub your shoe in the pile of dirt so as to coat the bottom with dust.

3. Place a white piece of paper on the floor and walk over it, making a 2-D footwear impression on the piece of paper (you may need to step across it diagonally).

4. Repeat steps 2 and 3 four more times, making five impressions. Make sure that adequate detail is found in each impression and that the contrast is sufficient for comparisons. Staple all impressions to your worksheet. Clean up the dirt on the floor.

5. **Record** your observations for the above process.

Part B Tape Footwear Impression

1. Obtain a roll of scotch tape. Pull off about 1–1.5 feet of tape. Tape about 1 inch of the end on the bench, letting the long end hang free. Repeat the process 5–6 more times overlapping each successive piece of tape making a sheet of tape large enough for the sole of your shoe. This step may not be required if sheets of adhesive are available.

2. Remove the tape sheet and place it, sticky side up, on the floor. Using the opposite shoe as the one used in part A, step on the sheet of tape only once (**do not** walk around).

3. While holding your foot up, remove the piece of tape. (**Do not** press the sheet of tape with your hand on your shoe after stepping on the tape. This can lead to a double impression.)

4. Place the sheet of tape on a piece of white paper with the sticky side down, fastening it to the paper. Staple it to your worksheet.

5. **Record** your observations for the above process.

Part C 2-D Footwear Impression Comparisons

1. Find the standard set of impressions made by the instructor using fingerprint ink. Compare all of the standard impressions and **record** in your worksheet three or

four points of natural variation (characteristics that **are not common to all** the standard prints).

2. Find the unknown impression created with dust or ink on a piece of paper. Identify three or four potential individual characteristics. Compare this impression to the standard impressions and determine if it was made by the same shoe. **Record** your observations.

Part D 3-D Footwear Cast (Two groups may need to come together for this part)

1. Obtain a pan large enough to make a 3-D footwear impression. Place about 1 inch of dirt in the pan and add a little water, making it slightly muddy. (The dirt must be moist, but not runny.)

2. Have one person in the group with the smallest shoes step into the mud in a walking motion. Rinse the dirty shoe in the sink.

3. Prepare a batch of plaster of Paris according to the instructions on the label. Make sure that you have enough to fill in the impression and have a cast of about 0.5 inches thick. If the impression is not deep enough, you may choose to prepare a small dam around it prior to applying the plaster of Paris.

4. Pour the plaster of Paris over the impression. One may want to pour the mixture next to the impression and allow it to flow into the impression; however, this may be more difficult.

5. Allow the cast to become slightly hard. Carve into the cast something to help you identify it for the next lab. **Remember to wear the same pair of shoes for the next lab.**

6. **Record** three observations for the above process.

Part E Physical Match with Paper

1. Cut a piece of paper into four equal pieces using scissors. Rip each piece of paper in the same direction two times across the shortest length. Mix up all 12 pieces of paper and give them to your partner. Do not use lined paper.

2. Prove to yourself that each rip and cut your partner made is unique and put the piece of paper back together. (Remember, you may need to turn the pieces over in order to get them to fit.)

Part F Physical Match with Sticks

1. Find the sticks broken by the instructor. Attempt to match the broken sticks. **Record** your results.

Part G Completion of 3-D Footwear Cast (This part will need to be finished during the next lab period.)

1. Remove the 3-D footwear impression from the pan.

2. Place the remaining dirt found in the pan back in the dirt bucket.

3. Rinse the impression under the sink, removing as much dirt and debris as possible. If you use a brush, be careful not to damage the cast.

4. Remove the shoe that made the impression from your foot. Compare the cast directly to the shoe. **Record** your observations.

5. Clean up when finished.

Name:_____   Date:_____

Experiment 20 Worksheet

Results and Observations

Part A Paper Footwear Impression

o   Observations

Part B Tape Footwear Impression

o   Observations

Part C 2-D Footwear Impression Comparisons

o   Draw a rough sketch of one of the known impressions, identifying and describing three or four possible points of natural variation.

o Draw a rough sketch of the known impression identifying and describing three or four individual characteristics.

o Was the unknown impression in part C made by the same shoe as the standard impressions? Support your answer with observations made above.

Part D 3-D Footwear Cast

o Observations

Part E Physical Match with Paper

o Discuss difficulties encountered and techniques employed when matching paper pieces.

Part F Physical Match with Sticks

o Discuss difficulties encountered and techniques employed when matching broken sticks.

o Which broken stick pieces matched? Explain.

Part G Completion of 3-D Footwear Cast

o Observations (**record** both class and individual characteristics)

Experiment 20 Questions

1. Compare the first five impressions made on paper in part A. Identify three or four individual characteristics by drawing lines and labeling. (Individual characteristics must be unique to the shoe and found in all five impressions.)

2. Identify three or four points on the above five impression that appear to be individual characteristics, but are not found consistently on all five impressions (most likely they are only found on one or two impressions). These points are due to the natural variation in which the print was made.

3. Discuss the importance of understanding the natural variation in the standard impressions prior to making a comparison to an unknown.

4. Pretend that the impression made with the tape sheet is a "lift" from a crime scene. Compare this "lift" to the five standard impressions made on paper. Discuss both class and individual characteristics. You know that a different shoe made the impression, but can you support this with individual characteristic differences? You may want to mark points on this impression. (The tape impression on the second shoe has been intentionally reversed so as to be compared to the standard impressions.)

Experiment 21
Microscopy II

Introduction

The microscope has been, and continues to be one of most powerful and versatile tools available to the forensic scientist. The microscope offers the analyst the ability to view fine detail that cannot be seen with the naked eye. In addition, the microscope can be used to measure dimensions and optical properties (e.g., refractive index and birefringence) of objects.

Despite the relatively simple understanding of the use of a microscope to magnify objects, the design and operation of such an instrument is highly sophisticated. Such sophistication is primarily due to the nature of the medium (i.e., light) allowing the microscope to operate. Light is a source of energy that has perplexed scientists for years, and continues to baffle many experts. However, despite the confusion, there are many aspects of light that we do understand. Such knowledge gives us the ability to use light to probe properties of virtually all types of matter. The microscope is simply one example of many instruments capable of manipulating light for the purpose of investigation.

Purpose

To become familiar with the operation of several types of microscopes (i.e., stereoscopic microscope, compound/biological microscope, polarized light microscope, and comparison microscope). Various types of samples will be examined requiring different sample preparation techniques. Students will also become familiar with the use of the microscope for physical match analyses and the measurement of dimensions.

Materials

Obtain the following materials:
1. Fibers (from your clothing)
2. Hair (from you scalp or body)

The following materials and chemicals will be supplied as needed:
1. Stereoscopic, compound/biological, polarized light, and comparison microscopes
2. Scissors
3. Three 1-mL disposable plastic pipettes
4. Color Desk Jet printout
5. Dead bugs
6. Microscope slides and covers
7. Permount^{TM}
8. Xylene
9. Pond water
10. Dimension measuring optics and slides
11. Spent cartridge casings or tool marks

Procedures

Part A Stereoscopic Microscope

1. Obtain three 1-mL plastic pipettes. Cut the tip off of the three pipettes in the same place (e.g., at the first or second partition).

2. Label the pipettes and the cut tips with a marker, each having a unique label (e.g., A–F). **Record** on a piece of paper which tip goes with each pipette (e.g., A and F, B and C, D and E). Mix up the tips and the pipettes and give them to your partner.

3. Using a stereoscopic microscope, attempt to perform a physical match of the tips and pipettes given to you by your partner. Check your results with your partner and **record** your observations.

4. Each partnership obtain a Desk Jet print out from the instructor. View the printout under the stereoscopic microscope. **Record** your observations.

5. Each partnership obtains a microscope slide, a piece of double-stick tape, and a dead insect. Place the piece of double-stick tape on the center of the microscope slide. Mount the bug on the tape. Examine the bug using a stereoscopic microscope. **Record** your observations.

Part B Polarized Light Microscope

1. Obtain a drop of pond water from the instructor and place it on a microscope slide. Place a cover slip over the drop of water. View it under the polarized light microscope using non-polarized light. **Record** your observations.

2. Cross the polars on the microscope (you may need to ask your instructor for assistance) and view the pond water sample. **Record** your observations.

3. Tease away some fibers from your clothes using your fingers or some tweezers. Add a drop of xylene to the microscope slide and place the fiber in the xylene. Orient the fiber such that it is found entirely on the slide. Add a drop of Permount™ to the slide and cover it with a cover slip. View the fiber using a polarized light microscope in both polarized and non-polarized light. Determine if the fiber is birefringent (it will appear to glow while the background is dark under crossed polars). **Record** your observations.

4. Pluck a hair from your scalp. Mount the hair in the same manner as the fiber. View the hair using a polarized light microscope in both polarized and non-polarized light. Determine if the hair is birefringent. **Record** your observations.

Part C Microscopic Measurements

1. Using the microscope equipped with an ocular having a measuring gradient, measure the diameter of the hair found on the microscope. This will require the calibration of the ocular gradient at a set magnification using a slide with measurement markings. You may need to have your instructor demonstrate. **Record** your results.

Part D Comparison Microscope (Optional)

1. View two impressions made using the same tool under the comparison microscope. **Record** your results.

2. Clean up when finished.

Experiment 21 Worksheet

Results and Observations

Part B Stereoscopic Microscope

- o **Record** observations and discuss difficulties and techniques developed matching cut pipettes.

- o Record observations of Desk Jet printout. Identify the three colors (excluding black) that compose the entire image. **Record** magnification used.

- o **Record** observations of the insect viewed under the stereoscopic microscope. Identify four features of the insect that are not apparent to the naked eye. **Record** magnification used.

Part C Polarized Light Microscope

- o Identify three features of the pond water that are not apparent to the naked eye under *non-polarized* light. **Record** magnification used.

o **Record** your observations of the pond water under *polarized* light. **Record** magnification used.

o **Record** your observations of the fiber under *non-polarized* light. **Record** magnification used.

o **Record** your observations of the fiber under *polarized* light. Is the fiber birefringent? **Record** magnification used.

o **Record** your observations of the hair under *non-polarized* light. **Record** magnification used.

o **Record** your observations of the hair under *polarized* light. Is the hair birefringent? **Record** magnification used.

Part D Microscopic Measurements

○ Draw a picture of the hair seen under the microscope. What is the diameter of the hair? **Record** magnification used.

Part E Comparison Microscope (Optional)

○ **Record** your observations.

Experiment 21 Questions

1. Based on your observations, discuss how a hair may be distinguished from a fiber.

2. Discuss all the steps required for determining microscopic dimensions as performed in lab. Would the apparent (i.e., observed) dimension be larger or smaller then the actual dimension if the magnification of the objective was increased without recalibration? Explain.

3. Theorize as to why the technique used in this laboratory for the measurement of microscopic dimensions is not perfectly accurate.

Experiment 22
Hair

Introduction

On average, 100 scalp hairs are shed from the human head daily. In addition, many crimes involving violent acts lead to forcibly removed hair. Such occurrences contribute to the relative abundance of hair at a crime scene. Hair is very robust and often remains unchanged for years after removal. Therefore, the majority of hair found at a crime scene, of course, may have nothing to do with the crime committed. In addition, due to the relatively miniscule nature of hair as evidence, the location of hair is frequently difficult to ascertain. And once found, the significance is often elusive even after a laboratory analysis.

Typically, hair analyses are performed by comparing several control hairs of a suspect or victim to those found at a crime scene. Comparisons are rarely definitive and hair is commonly considered class evidence. Individualization is only achieved through a positive nuclear DNA analysis of DNA extracted from adhering follicular tissue or hair root. It may also become important during an analysis to distinguish human hair from animal hair, requiring species identification.

Purpose

To become familiar with the microscopic characteristics of hair that allow for comparison. Students will also become familiar with the unique microscopic features of certain animal hairs (e.g., cow, dog, cat, deer, and mink) distinguishing them from human hairs.

Materials

Obtain the following materials:
1. Six scalp hairs (supply your own)

The following materials and chemicals will be supplied as needed:
1. Compound/biological, stereoscopic, and polarized light microscopes
2. Microscope slides
3. Microscope slide covers
4. Permount^{TM}
5. Xylene
6. Animal hairs (i.e., cow, dog, cat, deer, and rabbit)
7. Clear fingernail polish
8. Double-stick tape

Procedures

Part A Human Scalp Hair Comparison

1. Obtain six scalp hairs from your own head by either finding loose hairs or plucking. Preferably obtain samples from different locations on the scalp.

2. Mount three hairs on separate microscope slides using xylene and Permount^{TM} in the same fashion as described in experiment #23. Mount the other three samples using double-stick tape.

3. View each sample separately. Examine each hair for undulation (waviness), length, abnormalities, medullation, tip and root appearance, color treatment, and pigment distribution. All examinations may not require the use of the microscope (e.g., length). Examine the Permount^{TM} samples using the compound/biological or polarized light microscope and the double-stick tape samples using the stereoscopic microscope. **Record** your observations in your worksheet.

4. Find the individual in the lab who has hair most similar to yours. Exchange one Permount^{TM} slide and one double-stick tape slide. Compare these unknown samples to your reference samples. Can you distinguish them? Do they appear to share a common origin? **Record** your observations.

Part B Animal Hair Identification

1. Mount the following animal hairs separately using xylene and Permount^{TM} on individual microscope slides: cow, dog, cat, deer, and rabbit. Examine each under the compound/biological or polarized light microscope looking for characteristics distinct from human hairs (e.g., medullation, root appearance, etc.). If you have difficulty finding the medulla, use the polarized light microscope. The cortex is birefringent, while the medulla is not. The medulla typically appears black when using transmitted illumination. **Record** your observations.

2. Mount a deer hair sample on a microscope slide using double stick tape. Examine the hair under the stereoscopic microscope. **Record** your observations.

3. Make a scale cast of both a human hair and a mink hair using clear fingernail polish. Paint a thin coat of fingernail polish on the microscope slide. Carefully place the hair on the microscope slide, making sure it makes contact with the polish. Wait approximately 5–10 minutes for the polish to dry. Then carefully remove the hair. Examine under the compound/biological microscope. **Record** your observations.

4. Clean up when finished.

Name:_____   Date:_____

Experiment 22 Worksheet

Results and Observations

Part A Human Scalp Hair Comparison

o   Using the stereoscopic microscope, **record** and discuss the range of length, undulation, color, and any abnormalities found in your three scalp hairs (make sure you examine the full length of the hair). In a separate paragraph, discuss the natural variation of these features and the potential to use them in a comparison.

o Using the compound/biological or polarized light microscope, **record** and discuss the range of pigment distribution and medullation. In a separate paragraph, discuss the natural variation of these features and the potential to use them in a comparison.

o Using both microscopes, **record** and discuss the range of tip and root appearance. In a separate paragraph, discuss natural variation and the potential to use these features in a comparison.

o Discuss the microscopic comparison of a single hair from an individual with hair macroscopically similar in appearance to yours. Your discussion should include observations using both microscopes and whether or not hair properties of the samples fell within the natural variation of your hair. Could you distinguish the sample from your hair? Support your answer with observations made above.

Part B Animal Hair Identification

o Compare the macro- and microscopic similarities and differences between cat and rabbit hair.

o Compare the macro- and microscopic similarities and differences between dog and cow hair.

o   **Record** your observations for both deer preparations here. Discuss the unique characteristics of deer hair in comparison to the previous samples.

o   **Record** your scale cast observations for human and mink hair, describing the scale pattern of each. Discuss how you can distinguish between human and mink hair based on your scale cast observations.

Experiment 22 Questions

1. Based on your understanding of hair analyses, discuss what can be done with a single hair found at a crime scene if no reference standards are available.

2. Explain why hair is considered class evidence. Discuss when it might contain individual characteristics.

3. Theorize what potential problems may arise when attempting to identify a hair type (e.g., determine species). What other potential problems may arise when performing a hair comparison?

Experiment 23
Firearms

Introduction

Firearms are unfortunately used in a wide variety of crimes (murder, poaching, robbery, rape, etc.); and in each of these crimes, information associated with the firearm used can be found from discharged bullets and cartridge casings left at a crime scene. Sometimes firearm identification can be performed to a high degree of accuracy. Other times, however, only a general type of weapon may be ascertained from the evidence left. It is important that the analyst be familiar with all of the types of firearms and the markings left on bullets and cartridges for comparison.

Purpose

To introduce the student to firearms analyses. Each student will be introduced to various types of firearm ammunition, how to assemble and dissemble the ammunition, and how the spent ammunition can be used to potentially identify the firearm.

Materials

The following materials and chemicals will be supplied as needed:
1. Various powders and cartridge types
2. Bullet puller
3. 9-mm cartridge
4. Balance
5. Weigh boat
6. Calipers
7. Stereoscopic microscope
8. Bullets and cartridge casings fired from known rifles

Procedures

Part A Identification of Ammunition

1. View and identify items found in boxes 1–9. Make sure that you can identify each of the items found in each box. **Record** the identification of each item and many of its characteristics.

Part B Disassembling a Cartridge

1. Obtain a cartridge labeled #1 and a bullet puller. Weigh the cartridge and **record** your results in grams and grains (1 grain = 0.0648 grams). You will need to convert from grams to grains.

2. Insert the cartridge into the bullet puller and remove the bullet and powder from the casing by striking on the ground. Measure the mass of the bullet, casing, and powder separately and **record** your results in grams and grains. **Record** the appearance of the bullet and powder.

3. Reassemble the cartridge by placing the powder back into the case and twisting the bullet into the case. You may need to **carefully** hammer the bullet into the case (primer has been discharged). Please recover all powder for the next student. Return the cartridge.

4. Repeat steps 1–3 for cartridge #2.

Part C Comparing Bullet and Casing Markings

1. Obtain a box containing a bullet and a casing fired from a known gun. Determine the caliber by measuring the bullet diameter, count the number of land (scratched regions) and groove impressions, determine the direction of twist (scratch lines slant to right or left), and measure a range of width for the land and groove impressions (take at least three measurements at different locations). Measure from the bottom of the land impression. You may need a magnifying glass or a microscope. **Record** your results.

2. Identify any markings found on the side and/or bottom of the casing (e.g., potential ejector or extractor marks, breech markings, and firing pin markings). **Record** all manufacture markings.

3. Repeat step 1 and 2 for a second bullet and casing.

4. Clean up when finished.

Name:_____   Date:_____

Experiment 23 Worksheet

Results and Observations

Part A Identification of Ammunition

Item	Identification	Characteristics
1		
2		
3		
4		
5		
6		
7		
8		
9		

Part B Disassembling a Cartridge

	Grams #1	Grams #2	Grains #1	Grains #2
Cartridge				
Bullet				
Casing				
Powder				

o **Record** the appearance of the powder.

o **Record** the appearance of the bullet.

Part C Comparing Bullet and Casing Markings

o **Record** the bullet diameter, number of lands and grooves, range of land and groove widths, direction of twist, and general characteristics for bullet #1.

186

- o **Record** any identifiable ejector or extractor markings, breech markings, firing pin markings, manufacture markings, and general characteristics on casing #1.

- o **Record** the bullet diameter, number of lands and grooves, range of land and groove widths, direction of twist, and general characteristics for bullet #2.

- o **Record** any identifiable ejector or extractor markings, breech markings, firing pin markings, manufacture markings, and general characteristics on casing #2.

Experiment 23 Questions

1. Potential gun type can be determined by examining bullet diameter, number of lands and grooves, direction of twist, and land and groove impression widths. How might one be able to definitively conclude which gun fired the bullet found at a crime scene? Explain.

2. Based on your understanding of a basic firearms analysis, what precautions need to be followed when collecting firearms evidence such as bullets and casings so not to destroy the evidence?

3. Which, if any, of the firearms analysis techniques used in this laboratory can be performed on a partially mutilated bullet (i.e., it has partially mushroomed and one side is smashed)? Why or why not?

Experiment 24
Questioned Documents

Introduction

A questioned document can be any object containing handwritten or typewritten markings whose source or authenticity is in doubt. Documents may include, but are not limited to, checks, contracts, wills, letters, passports, licenses, lottery tickets, and petitions. A handwritten questioned document may be analyzed for more than just handwriting. Indentions, erasures, obliterations, alterations, pen inks, watermarks, paper fibers, and abnormalities may all be examined. Each analysis will typically require different techniques. A technique for the analysis of pen inks was performed in experiment 4. Techniques for the analysis of indentions, obliterations, and handwriting will be performed in this experiment.

Purpose

To introduce the student to a limited number of questioned document analysis techniques. Each student will become familiar with handwriting characteristics, indentions, and obliterations.

Materials

Obtain the following materials:
1. Seven sheets of paper
2. Pencil

The following materials and chemicals will be supplied as needed:
1. Unknown obliterations
2. Ultraviolet light source

Procedures

Part A Preparing Handwriting Exemplars

1. Using your writing hand, write the following in your normal handwriting

   Anne and Angus left for school two weeks ago. Anne has struggled living alone and Angus loves it. Each earn $16,661 a year working at Bob & Bobbie's Burgers. 45% of their income goes toward school!

2. Grasp your pen or pencil, making a fist with the point of the pen or pencil closest to your pinky. Write the above three sentences again.

3. Now hold the pen or pencil at your elbow between your forearm and bicep and write at least the first sentence again.

4. Write the three sentences one more time using your writing hand, but try to disguise your normal handwriting. Do not take an excessive amount of time. Try to write the disguised version in approximately the same amount of time it took to write the original in step 1. Answer questions 1–3. Turn in your exemplars with your worksheet.

Part B Preparing Signature Exemplars

1. Write your full signature ten times while sitting comfortably.

2. Write your full signature five times while standing.

3. Write your full signature five times, trying to disguise it. Again, try not to take an excessive amount of time writing your disguised exemplars. Answer questions 4–6. Turn in your exemplars with your worksheet.

Part C Indentions

1. Obtain at least five sheets of paper. Place them flat on a hard surface. On the top sheet, with the other four underneath, write your complete address, pressing moderately with a pen.

2. Exchange the sheet immediately under the first sheet with a partner. Be sure that he/she did not see your written address. **Record** the name of the person on the first sheet and turn it in with your worksheet.

3. Without seeing what your partner wrote on the first sheet, look closely for the indentions made on the sheet he/she gave you using oblique lighting. This will require you to tilt the paper at various angles, casting shadows on the indentions. Attempt to decipher the indentions. **Record** your results.

4. Next, using a sharpened pencil or pencil lead, very carefully and lightly highlight the region containing the indentions with the flat edge of the pencil tip. This process should darken the region around the indention, leaving the indention white. The increased contrast should aid in deciphering the address. **Record** your results, being sure to **record** the address. Turn in the developed indentions.

Part D Obliterations

1. Two sheets of paper are found having writing that has been obliterated or crossed out with a second writing utensil.

2. First, examine each sheet closely while placing them flat on a table. **Record** your results.

3. Second, examine each sheet using oblique lighting by holding each at various angles. **Record** your results.

4. Third, examine each sheet by holding them up to the light or an open window. **Record** your results.

5. Lastly, examine each sheet under ultraviolet light. **Record** your results, being certain to record everything developed.

6. Clean up when finished.

Experiment 24 Worksheet

Results and Observations

Part C Indentions

- o   Observations of indentions using oblique lighting

- o   Observations of indentions using pencil highlighting

Part D Obliterations

- o   Observations of sheets on a flat surface

- o   Observations of sheets using oblique lighting

o   Observations of sheets held up to the light

o   Observations of sheets using ultraviolet lighting

o   What was written under the obliterated writing on each page?

Experiment 24 Questions

1. The nature of someone's handwriting is recorded in the brain early during handwriting development and changes little after the high school years. Because handwriting is recorded in the brain, common patterns should be seen in sentences written using various writing utensil placements (e.g., hand, elbow, etc.). Carefully examine what was written in steps 1–3 of part A and prove this theory by finding at least five common patterns. Circle them and discuss them below. Things to examine include spacing, slant of letters, connecting strokes, pen lifts, unusual letter formations, beginning and ending strokes, base line slant, flourishes or embellishments, placement of i dots and t crossings, and general letter and word appearances.

2. Also, because handwriting style is recorded in the brain, many attempts to disguise handwriting are unsuccessful due to the many subconscious movements of the writing utensil when forming letters and words. Find at least five similar patterns between what was written in step 1 and step 4 of part A. Put a box around them and discuss them below.

3. Many exemplars are acquired by forcing the suspect to write a certain script under controlled conditions. Discuss the significance of making sure that an excessive amount of time is not taken when the suspect prepares an exemplar.

4. Nobody can write their signature exactly the same every time. A good indicator of forgery is finding a signature that looks exactly like an original. Find five differences in your ten signatures that are not common to all ten signatures. Circle them and discuss them below.

5. It is known that we do not sign our name exactly the same under all circumstances. A credit card receipt signature on a cold morning may look quite a bit different from the signature on a legal document. It is important when acquiring exemplars that conditions are as similar as possible. See if you can find two or three differences between the signatures you made standing up compared with those made sitting down. Put a box around them and discuss them below.

6. Also, because handwriting style is recorded in the brain, many attempts to disguise signatures are unsuccessful. Find at least five similar patterns between your normal and disguised signatures. Circle them and discuss them below.

7. Discuss what made the various lighting techniques successful in deciphering the obliterated writing. Why was contrast developed?